Digital Technology and Justice

Justice apps – mobile and web-based programmes that can assist individuals with legal tasks – are being produced, improved, and accessed at an unprecedented rate. These technologies have the potential to reshape the justice system, improve access to justice, and demystify legal institutions. Using artificial intelligence techniques, apps can even facilitate the resolution of common legal disputes.

However, these opportunities must be assessed in light of the many challenges associated with app use in the justice sector. These relate to the digital divide and other accessibility issues; the ethical challenges raised by the dehumanisation of legal processes; and various privacy, security, and confidentiality risks.

Surveying the landscape of this emergent industry, this book explores the objectives, opportunities, and challenges presented by apps across all areas of the justice sector. Detailed consideration is also given to the use of justice apps in specific legal contexts, including the family law and criminal law sectors. The first book to engage with justice apps, this book will appeal to a wide range of legal scholars, students, practitioners, and policy-makers.

Tania Sourdin is the Dean of the University of Newcastle Law School and was previously the Foundation Chair and Director of the Australian Centre for Justice Innovation (ACJI) at Monash University in Australia as well as the author of more than 120 books, articles, and chapters that focus on justice innovation and reform. In the past two decades, she has conducted extensive research into aspects of the dispute resolution and justice system which has included building and reviewing technologically enhanced systems. Parts of this book draw on discussion in Tania Sourdin, *Judges, Technology and AI*, forthcoming, (2021) (Edward Elgar).

Jacqueline Meredith is a Sessional Academic at Newcastle Law School and a PhD candidate at Melbourne Law School. She holds a Bachelor of Laws (Hons 1, University Medal) from the University of Newcastle and a Bachelor of Civil Law (Dist) from the University of Oxford. In addition to work carried out on various research projects exploring the intersection of law and technology, Jacqueline has also published in the areas of labour law, medical law and ethics, and legal education. She has previously worked as a Judge's Associate in the Supreme Court of Victoria and a Senior Researcher and Associate Lecturer in Law at the University of Newcastle.

Bin Li is Lecturer at the University of Newcastle Law School. Before moving to Australia in 2016, he worked as an Associate Professor of Law at Beihang University in China, previously known as Beijing University of Aeronautics and Astronautics (BUAA), or Beihang University, in China. His research interests include the technological impact on dispute resolution processes and access to justice issues.

Digital Technology and Justice

Justice Apps

**Tania Sourdin,
Jacqueline Meredith,
and Bin Li**

Routledge
Taylor & Francis Group

LONDON AND NEW YORK

First published 2020
by Routledge
2 Park Square, Milton Park, Abingdon, Oxon OX14 4RN

and by Routledge
605 Third Avenue, New York, NY 10017

First issued in paperback 2022

Routledge is an imprint of the Taylor & Francis Group, an informa business

Publisher's Note
The publisher has gone to great lengths to ensure the quality of this
reprint but points out that some imperfections in the original copies may
be apparent.

British Library Cataloguing-in-Publication Data
A catalogue record for this book is available from the British Library

Library of Congress Cataloging-in-Publication Data
A catalog record has been requested for this book

ISBN 13: 978-0-367-65018-6 (pbk)
ISBN 13: 978-0-367-62352-4 (hbk)
ISBN 13: 978-1-00-312703-1 (ebk)

DOI: 10.4324/9781003127031

Typeset in Times New Roman
by codeMantra

Contents

Preface

This book is the result of research conducted at the Newcastle Law School, the University of Newcastle, Australia, that has taken place under the auspices of a research cluster that focusses on justice innovation based on digital technology. As part of the research undertaken through this research cluster, a range of matters relating to technology, mobile applications ('apps'), human-centred design, ethics, case management, and improved justice processes are explored.

Through the research cluster, the authors have also conducted more specific research into 'justice apps'. These are mobile and web-based resources intended to assist individuals with tasks in the justice area.[1] Justice apps can target lawyers, the general public, and others. Such apps can promote more efficient legal service delivery.[2] Apps targeting the general public – sometimes termed 'DTP' (direct-to-public) apps – can also make legal services easier to access and allow users to engage with self-help processes.[3] More sophisticated 'chatbot' or 'robolawyer' apps can offer recommendations or solutions based on conditional and causal decision logic trees, and, in some cases, more advanced artificial intelligence ('AI')[4] techniques.[5] Newcastle Law School has also fostered the development of mobile apps, such as *Bernie*, which is focussed on supporting domestic violence defendants to strengthen their positive decision-making around behaviours, and the *Know the Law* app, which is delivering legal information to students who have encountered legal issues, particularly surrounding tenancy.

Initial research into justice apps was supported by an extensive literature review undertaken to evaluate an app called the *Adieu* system, which is focussed on family law disputes in Australia and which resulted in the authors considering the research into apps in the justice sector and elsewhere. Professor Sourdin was engaged to evaluate the system and, as a result, reviewed research, considered user demographics, surveyed users, and interviewed professionals involved in the system. The *Adieu* system also changed during that evaluation programme and, in addition to developments relating to '*Lumi*' – a family law robot – that were already well underway, *Adieu* grew to incorporate an additional AI system that was focussed on supporting the exchange of financial information between divorcing couples and analysing that information. Some of the findings made in relation to the *Adieu* system appear in this book, although much of the content is related to justice apps that operate around the world and which can be used by lawyers, litigants, dispute resolution practitioners, judges, and other experts.

The authors thank *Adieu* and in particular Bill and Andrew Wight, who not only supported the research that led to this book but also supported the evaluation of their system. In sponsoring this research, they contributed to the future development of apps in the justice sector and enabled additional research into the issues that surface in relation to the justice system and technology.

The authors also thank the Faculty of Business and Law at the University of Newcastle, Australia, which supported an extension of the initial research project that has enabled them to undertake additional work on related international developments that is reflected throughout this book.

During the writing of this book, the COVID-19 pandemic resulted in massive changes to justice systems around the world. Such changes meant that many apps were used much more extensively and often for the first time in parts of the justice system. In particular, video conferencing apps reshaped the way in which many people involved in the justice sector worked. Apps have, however, led to other extensive changes in the sector over the last five years in terms of how people understand the law, engage with the justice system, and achieve outcomes. The authors predict that over the next five years, apps will continue to reshape the justice system, perhaps more quickly, as a result of the changes made necessary by the global pandemic. In the meantime, they argue that, in developing and using justice apps, caution should be taken by relevant decision-makers to address a wide range of legal and ethical issues with a view to ensuring 'justice' for users.

Notes

1 Teresa Scassa et al., 'Developing Privacy Best Practices for Direct-to-Public Legal Apps: Observations and Lessons Learned' (2020) 18(1) *Canadian Journal of Law and Technology* (forthcoming).

2 Jena McGill, Suzanne Bouclin and Amy Salyzyn, 'Mobile and Web-based Legal Apps: Opportunities, Risks and Information Gaps' (2017) 15 *Canadian Journal of Law and Technology* 229, 238; Teresa Scassa et al., 'Developing Privacy Best Practices for Direct-to-Public Legal Apps: Observations and Lessons Learned' (2020) 18(1) *Canadian Journal of Law and Technology* (forthcoming).

3 Jena McGill, Suzanne Bouclin and Amy Salyzyn, 'Mobile and Web-based Legal Apps: Opportunities, Risks and Information Gaps' (2017) 15 *Canadian Journal of Law and Technology* 229, 239–40.

4 See Margaret A. Boden, *Artificial Intelligence: A Very Short Introduction* (Oxford: Oxford University Press, 2018), 1, who notes that AI is comprised of a variety of dimensions of information-processing that may include 'perception, association, prediction, planning, motor control', previously associated with humans.

5 Judith Bennett et al., 'Current State of Automated Legal Advice Tools' (Discussion Paper No 1, The University of Melbourne, April 2018) 26.

1 Introduction

Justice apps and justice

Over the past decade, there has been considerable growth in internet connectivity, online dispute resolution ('ODR'), and the production and improvement of justice apps.[1] Justice apps are mobile and web-based resources intended to assist individuals with legal tasks.[2] In recent years, apps have become an increasingly popular way of accessing information and connecting to justice services. In 2017, there were almost 200 billion mobile app downloads,[3] although only a small proportion of these apps related to the justice sector.[4] Since 2017 there has, however, been a significant increase in the range and type of justice apps which can be oriented towards lawyers and other experts, as well as the general public.

Justice apps can make legal services easier to access, guide users through legal choices, and allow them to engage with self-help processes. More sophisticated legal 'chatbot' or 'robolawyer' apps can offer recommendations based on complex artificial intelligence (AI) techniques. This book's primary aim is to explore the objectives, opportunities, and limitations of justice apps. Whilst the focus will be on justice apps broadly, detailed consideration is also given to their use in specific legal contexts, including the family law and criminal law sectors.

As early as 2012, in the analysis paper, *Harnessing the Benefits of Technology to Improve Access to Justice*, the Australian Government noted that:

> The development of mobile software applications and optimised websites for use on mobile phones is increasingly becoming a necessity if an agency or organization wishes to have a strong online presence and increase their reach to the public, especially people in RRR areas which is very important for legal assistance. Given

the current use of mobile media devices and the projected increase in the use of such devices to access information in the future, there will be an increase in expectations from the public that they will be able to access information through these devices and that this information will be current.[5]

In the years since 2012, justice apps have been developed, evolved, and extended to enable many people to access information about the justice system and, under some circumstances, support levels of engagement that were not contemplated a decade ago. At the same time, there has been little literature or research relating to these developments, with no research specifically focussed on justice apps. This book is directed at that omission and the authors note that such apps have the potential to reshape the justice sector in a number of critical ways. Given the proliferation of personal media devices and significant advances in technology over the past decade, the production, improvement, and uptake of justice apps will only continue to expand.

This book considers developments in a range of diverse jurisdictions, including Australia, the United States of America (the U.S.), and China. Whilst this is a rapidly developing area, the book's focus is on the underling objectives, opportunities, and challenges presented by justice apps, meaning that the frameworks and approaches that are developed are of ongoing relevance into the future. In addition, the use of some case study material supports consideration of the issues that are likely to emerge in the justice app area over the coming decade.

The context within which justice apps are used

In general, and to place the emergence of justice apps within some context, Sourdin[6] has previously noted that changing and emerging technologies have considerable relevance to the continuing evolution of alternative dispute resolution (ADR) processes and the justice system in general.[7] Sourdin's taxonomy suggests that there are three main and interlinked ways in which technology is reshaping justice systems, and justice apps can also be considered and analysed by using this taxonomy. First, and at the most basic level, technology can assist to inform, support, and advise people involved in justice activities ('supportive technologies'). Second, technology can replace activities and functions that were previously carried out by humans ('replacement technologies'). Finally, at a third level, technology can provide for very different forms of justice, particularly where processes change significantly ('disruptive technologies').[8] Justice apps can fall within any of these three categories. Some apps can involve both 'supportive' and 'replacement' technologies, whilst others may have aspects that are 'disruptive'.

Justice apps can target lawyers, the general public, and others.[9] Apps targeting lawyers often promote more efficient legal service delivery and can assist to streamline legal research.[10] Apps targeting the general public – sometimes termed 'DTP' (direct-to-public) apps[11] – can make legal services easier to access and allow users to engage with self-help processes.[12] Specifically, justice apps can help users identify if there is a legal issue, guide users through legal choices, assist with the drafting and filing of legal documents, and provide referrals to legal service providers.[13] More sophisticated legal 'chatbot' or 'robolawyer' apps can offer recommendations or solutions based on conditional and causal decision logic trees, and in some cases, more advanced AI techniques.[14]

Although there is currently very limited research on the use of apps in the legal context, there is a substantial amount of domestic and international research exploring the use of technology in the justice sector more broadly. This includes research on the benefits and risks associated with ODR processes. More recent scholarship has also considered the impact of automation on governmental and administrative decision-making.[15]

Growth in apps and connectivity

As noted previously, the production and improvement of justice apps is an area of significant recent development. This is partly because apps may support those who are comfortable with smartphones and mobile devices but intimidated by, or unable to access, computers.[16] Smartphones that enable apps are a relatively recent invention. Although some basic smartphones with limited operation were in use in the early part of the 21st century, it is only over the past decade that smartphones, which enable users to download or purchase apps from a centralised facility, make payments, use cloud storage and synchronise updates and app functions, have been adopted by a growing proportion of people in developed and developing countries.

Indeed, in 2020, mobile phones have become the dominant means by which people access the internet. For example, in terms of internet connectivity, the Australian Bureau of Statistics (ABS) has noted that in 2016–2017, 87 percent of persons were internet users, with mobile or smartphones used by 91 percent of connected households.[17] In 2018, there were approximately 27 million mobile handset subscribers in Australia.[18] The number of mobile handset subscribers, and the volume of data they download with these devices, continues to grow. Globally, it is estimated that 62.6 percent of the world's population, or 4.78 billion people, are mobile phones users in 2020.[19]

Much app development resulted from the establishment of third generation ('3G') mobile communication technologies in the early 2000s.[20] 3G development made smartphone technology possible, and consumers have increasingly made the switch. A 2018 report predicted that the rise in affordable data plans and the establishment of fifth generation (5G) network in Australia would result in the uptake of mobile devices as a primary entertainment (or content) device.[21]

Worldwide data on app usage supports this conclusion. In 2017, as noted previously, there were almost 200 billion mobile app downloads, an increase of 32 percent from the prior year.[22] While app usage has increased as a result of this influx of new apps, more surprising is the way apps have changed the technological appetite of consumers.[23] Consumers are spending more time on apps than on any other media form. A 2017 survey of American consumers found that 50 percent of all digital media usage time was spent on apps, compared to 34 percent on desktops.[24] This is up 6 percent from 2015.[25] On average, consumers spent 2.3 hours a day on apps, with 18- to 24-year-olds being the most frequent users.[26]

In the five years from 2015 to 2020, there has also been a massive growth in justice apps of varying levels of sophistication. Some of these have been linked to court or tribunal systems and are often still being developed or trialled (e.g., *Intelligent Terminal Handling* App in Shanghai[27]; *eBRAM* in Hong Kong[28]; *Guided Resolution* in the New South Wales (NSW) Civil and Administrative Tribunal[29]), whilst others have been targeted at clients or lawyers (e.g., *Rocket Lawyer*[30]; *Ailira*[31]).

Despite this growth in smartphone use and the development and uptake of apps, issues have been raised about a digital divide and digital inclusiveness. For instance, in the United Kingdom (the UK) in 2013, 78 percent of the population reported using the Internet. However, in a 2018 report, it was noted that the scale of digital exclusion in the justice context specifically was unclear.[32] In China, as of March 2020, the number of mobile internet users has reached 904 million, including 255 million in rural areas, and 649 million from urban regions. In terms of age demographics, those aged between 20 and 29 years constituted the highest proportion of users (21.5 percent), closely followed by those aged between 30 and 39 years (20.8 percent). Notably, users aged 50 years and older had more limited usage at 16.9 percent, suggesting more limited mobile internet use by a range of people within some populations.[33]

There are continuing issues surrounding the digital divide that can be related to age, data access (connectivity), language, and education. These issues are discussed in Chapter 5 of this book and have specific implications for the development and use of justice apps.

Growth in justice apps

Within the justice sector, apps have become an increasingly popular way of accessing information and connecting to services. In November 2019, a search by the authors of the Australian *Apple* and *Google Play* stores for legal or justice-oriented apps revealed approximately 30 apps directed at Australian users. The vast majority of these fell within the category of supportive technologies,[34] which provide free access to legal information. The most basic of these apps allow users to view court lists[35] or search the legal databases for court decisions.[36]

Other apps are educational in nature, providing information about certain areas of the law. An example of this is the *LegalAidSA* app created by Legal Aid South Australia. The app is free to download and provides legal information and resources on a wide range of areas, including family law, wills, and motor vehicle accidents. The app also facilitates calls to free legal help telephone lines and provides travel directions to South Australian Legal Aid offices.[37] Similarly, the *NSW Pocket Lawyer* app provides information about common criminal and traffic offences, a defendant's legal rights, and other information to assist defendants with court processes and procedures.[38]

The Women's Legal Service Queensland has also created a free app – *PENDA* – which aims to empower victims of domestic and family violence with access to legal, financial, and safety information.[39] A June 2018 report evaluating the app identified good user engagement, with users returning to the app and interacting with the links and information provided.[40] Supportive justice apps can also connect individuals with legal professionals. An example is *SeekVisa* which connects users with migration agents, lawyers, advisers, and/or consultants throughout Australia.[41]

Several information and service apps have also emerged outside Australia. In the U.S., for example, *Ask a Lawyer: Legal Help* enables people to message and chat with lawyers for free.[42] Another U.S. app, *BernieSez*, allows consumers to upload a photograph of a ticket or other charge-related paperwork, with lawyers then able to compete to win work from the consumer.[43] Further developments in the app market have seen a shift from providing general advice, to assisting people collect evidence or otherwise prepare for a trial. For instance, the *Florida Courts HELP* app assists self-represented litigants in family law cases by providing access to hundreds of forms that can be completed on the app.[44]

There has also been growth in legal 'chatbot' or 'robolawyer' apps which offer legal advice based on conditional and causal decision logic trees, and in some cases, more sophisticated AI techniques.[45] *DoNot-Pay*, created in September 2015, is an AI chatbot app,[46] which, as of June 2016, had helped over 250,000 people challenge traffic and parking tickets across the UK and the U.S., with a 40 percent success rate.[47]

Another AI chatbot, *LISA*, was created in the UK in 2017 and allows parties to draft legally binding non-disclosure agreements.[48] *LISA* has also launched a number of property contract tools and is expected to branch into other areas as it grows.[49] In addition to assisting individuals who are seeking to navigate the legal system on their own, AI chatbot apps can also connect individuals with legal service providers. *Gideon* is 'an AI-powered intake, triage, and referral platform', based in the U.S., which is designed to 'route new clients to the right lawyer' and 'create new client matters'.[50] Using a proprietary, predictive AI system, *Gideon* learns a law firm's preferences on particular case types and predicts case outcomes, expenses, and estimated staff utilisation.

Conclusions

The above discussion reveals a significant variation in justice apps, within and across jurisdictions. Returning to the original taxonomy discussed at the beginning of this chapter, it seems clear that most early justice apps were supportive. That is, they provided information and support to a range of users; however, they were passive in that they did not support much user interaction. Apps of this nature may therefore raise fewer security or privacy concerns (see discussion in Chapters 3 and 5 of this book). Over the past five years, there has been a rapid development in apps that also rely on replacement technologies. A key feature of these apps is that functions previously undertaken by human researchers, case managers, and others, are replaced by technologies that are available through the operation of the app.

Apps can also be used by those within the justice sector to assist in decision-making or at other levels to support evidence gathering. Whilst the authors of this book note how important the latter apps can be and acknowledge that they raise a unique set of issues, including concerns about surveillance and tracking, the focus in this book is on those justice apps that may play a more critical role in decision-making. This is not to suggest that issues with other apps should remain unexplored; however, questions relating to evidence about their effectiveness would be best addressed in another work. Some more recent justice apps could be categorised as 'disruptive', particularly where they are directed at significantly changing the way that work takes place and creating very different ways of working. Some of these more disruptive developments are discussed later in this book, with particular reference to the use of more developed AI approaches that can support technologies that may completely change the way in which justice sector interactions take place. This may occur, for instance, through the replacement of human interactions with interactions via a chatbot with a virtual assistant. Additionally, human experts may no longer be engaged to provide expert information in a legal dispute or a transactional setting, such as where financial expert input is replaced by a form of AI.

Notes

1 An app is a software application that is most often used to refer to a mobile app that runs on a phone or portable device and can also include a software application that runs on a desktop computer or website.
2 Teresa Scassa et al., 'Developing Privacy Best Practices for Direct-to-Public Legal Apps: Observations and Lessons Learned' (2020) 18(1) *Canadian Journal of Law and Technology* (forthcoming); Ottawa Faculty of Law Working Paper No. 2019-35. Available at SSRN: https://ssrn.com/abstract=3464400 or doi:10.2139/ssrn.3464400.
3 Kungpo Tao and Paulette Edmunds, 'Mobile APPs and Global Markets' (2018) 8 *Theoretical Economics Letters* 1510, 1511.
4 It has been noted that by 2020, 'Of the estimated 4.5 million apps available in the Google and Apple app stores, a million collectively pertain to health, fitness, nutrition, and well-being in general': See Clarence Baxter, Julie – Anne Carroll, Brendan Keogh, Corneel Vandelanotte, 'Assessment of Mobile Health Apps Using Built-In Smartphone Sensors for Diagnosis and Treatment: Systematic Survey of Apps Listed in International Curated Health App Libraries' (2020) 8(2) *JMIR Mhealth Uhealth* 2.
5 See Australian Government, *Harnessing the Benefits of Technology to Improve Access to Justice* (2012) <https://webarchive.nla.gov.au/awa/20140212011730/http://www.sclj.gov.au/agdbasev7wr/sclj/harnessing_the_power_of_technology_analysis_paper.pdf>.
6 This material is drawn from and discussed in more detail in Tania Sourdin, 'Justice and Technological Innovation' (2015) 25 *Journal of Judicial Administration* 96, 105. This taxonomy is further discussed in: Tania Sourdin, Bin Li and Tony Burke, 'Just, Quick and Cheap? Civil Dispute Resolution and Technology' (2019) 19 *Macquarie Law Journal* 17, 19; Tania Sourdin, 'Judge v Robot: Artificial Intelligence and Judicial Decision making' (2018) 41(4) *University of New South Wales Law Journal* 1114, 1118; See also Tania Sourdin, *Judges, Technology and AI*, forthcoming (2021) (Edward Elgar).

7 See also Stefan Lancy, 'ADR and Technology' (2016) 27 *Australasian Dispute Resolution Journal* 168.

8 This material is drawn from and discussed in more detail in Tania Sourdin, 'Justice and Technological Innovation' (2015) 25 *Journal of Judicial Administration* 96, 105.

9 Jena McGill, Suzanne Bouclin and Amy Salyzyn, 'Mobile and Web-based Legal Apps: Opportunities, Risks and Information Gaps' (2017) 15 *Canadian Journal of Law and Technology* 229, 237.

10 Jena McGill, Suzanne Bouclin and Amy Salyzyn, 'Mobile and Web-based Legal Apps: Opportunities, Risks and Information Gaps' (2017) 15 *Canadian Journal of Law and Technology* 229, 238.

11 Teresa Scassa et al., 'Developing Privacy Best Practices for Direct-to-Public Legal Apps: Observations and Lessons Learned' (2020) 18(1) *Canadian Journal of Law and Technology* (forthcoming; Ottawa Faculty of Law Working Paper No. 2019-35. Available at SSRN: https://ssrn.com/abstract=3464400 or doi:10.2139/ssrn.3464400).

12 Jena McGill, Suzanne Bouclin and Amy Salyzyn, 'Mobile and Web-based Legal Apps: Opportunities, Risks and Information Gaps' (2017) 15 *Canadian Journal of Law and Technology* 229, 239-40.

13 See generally Sherley Cruz, 'Coding for Cultural Competency: Expanding Access to Justice with Technology' (2019) 86 *Tennessee Law Review* 347, 361.

14 Judith Bennett et al., 'Current State of Automated Legal Advice Tools' (Discussion Paper No 1, The University of Melbourne, April 2018) 26. See also Sherley Cruz, 'Coding for Cultural Competency: Expanding Access to Justice with Technology' (2019) 86 *Tennessee Law Review* 347, 364.

15 See, eg, Monika Zalnieriute, Lyria Bennett Moses and George Williams, 'The Rule of Law and Automation of Government Decision-Making' (2019) 82(3) *Modern Law Review* 425, 426-7.

16 See Joe Dysart, '20 Apps to Help Provide Easer Access to Legal Help' (2015) *American Bar Association Journal* <http://www.abajournal.com/magazine/article/20_apps_providing_easier_access_to_legal_help>.

17 Australian Bureau of Statistics, *Household Use of Information Technology, Australia, 2016-17* (Catalogue No 8146.0, 28 March 2018).

18 Australian Bureau of Statistics, *Internet Activity, Australia, June 2018* (Catalogue No 8153.0, 2 October 2018).

19 Jasmine Enberg, 'Global Mobile Landscape: A Country-by-Country Look at Mobile Phone and Smartphone Usage 2016' (Emarketer, 2016).

20 Anuj Pal Kapoo and Madhu Vij, 'How to Boost your App Store Rating? An Empirical Assessment of Ratings for Mobile Banking Apps' (2020) 15(1) *Journal of Theoretical and Applied Electronic Commerce Research* 99, 99.

21 Deloitte, *Behaviour Unlimited Mobile Consumer Survey 2018: The Australian Cut* (Report, 2018) 5.

22 Kungpo Tao and Paulette Edmunds, 'Mobile APPs and Global Markets' (2018) 8 *Theoretical Economics Letters* 1510, 1511.

23 Comscore, *The 2017 US Mobile App Report* (Report, 2017) 7. Cf Comscore, *The 2015 US Mobile App Report* (Report, 2015) 6.

24 Comscore, *The 2017 US Mobile App Report* (Report, 2017).

25 Comscore, *The 2015 US Mobile App Report* (Report, 2015) 6.

26 Comscore, *The 2017 US Mobile App Report* (Report, 2017) 7.

27 Yadong Cui, *'Artificial Intelligence' Makes the Court System More Just, Efficient and Authoritative* <https://law.stanford.edu/china-law-and-policy-association-clpa/articles/>.

28 eBRAM, *eBRAM* <http://ebram.org/>.

29 Guided Resolution, *Guided Resolution* <https://www.guidedresolution.com/company/news>.

30 Rocket Lawyer, *Rocket Lawyer* (2019) <https://www.rocketlawyer.com/>.

31 Ailira, *Ailira* <https://www.ailira.com/>.

32 Richard Susskind, *Online Courts and The Future of Justice* (Oxford University Press, 2020), 216.

33 Cyberspace Administration of China, *The 45th China Statistical Report on Internet Development* (2020), 20–5 <http://www.cac.gov.cn/2020-04/27/c_1589535470378587.htm>.

34 See generally Tania Sourdin, 'Justice and Technological Innovation' (2015) 25 *Journal of Judicial Administration* 96, 105.

35 See, eg, NSW Department of Attorney General and Justice, *Search NSW Court Lists* (2017) <https://play.google.com/store/apps/details?id=au.gov.nsw.agd.onlineregistry.courtlistsearch&hl=en>.

36 See, eg, AustLII Foundation, *AustLII* (2015) <https://play.google.com/store/apps/details?id=au.edu.austlii.mobile.android&hl=en>.

37 Legal Services Commission of South Australia, *LegalAidSA* (2019) <https://play.google.com/store/apps/details?id=com.andromo.dev86688.app95890>.

38 Sydney Criminal Lawyers, *NSW Pocket Lawyer* (2014) <https://apps.apple.com/au/app/nsw-pocket-lawyer/id918902178>.

39 Women's Legal Service Queensland, *PENDA* (2017) <https://penda-app.com>.

40 Ithaca Group, *Evaluation of PENDA: A Financial Empowerment App for Women Escaping Domestic and Family Violence* (Final Report, June 2018) 16.

41 SeekVisa, *Australian Visa & Immigration: SeekVisa* (2019) <https://play.google.com/store/apps/details?id=com.skycap.seekvisa>.

42 Lawyer Apps LLC, *Ask a Lawyer: Legal Help* (2015) <https://ask-a-lawyer-legal-help.soft112.com/>.

43 Legal Software Solutions LLC, *BernieSez* (2020) <https://www.berniesez.com/>.

44 Florida Courts, *Florida Courts HELP* (2020) <https://help.flcourts.org/download-the-app/>.

45 Judith Bennett et al., 'Current State of Automated Legal Advice Tools' (Discussion Paper No 1, The University of Melbourne, April 2018) 26. See also Sherley Cruz, 'Coding for Cultural Competency: Expanding Access to Justice with Technology' (2019) 86 *Tennessee Law Review* 347, 364.

46 Donotpay, *DoNotPay* (2020) <https://donotpay.com/>.

47 Judith Bennett et al., 'Current State of Automated Legal Advice Tools' (Discussion Paper No 1, The University of Melbourne, April 2018) 38.

48 Robot Lawyer Lisa, *LISA* (2020) <https://robotlawyerlisa.com/>; Judith Bennett et al., 'Current State of Automated Legal Advice Tools' (Discussion Paper No 1, The University of Melbourne, April 2018) 38.

49 Caroline Hill, 'Trending: Robot Lawyer LISA Launches Suite of Property Contracts', *Legal IT Insider* (6 November 2017) <https://legaltechnology.com/latest-news/robot-lawyer-lisa-launches-suite-of-property-contracts/>.

50 Gideon, *Gideon* <https://www.gideon.legal/>.

2 Digital technology use in the justice sector

Introduction

Within the justice sector, a vast array of very different technological systems and processes operate. For example, courts may use case management systems to assist with court workflow that may have automatic listing and other features. Often such systems operate using what is known as 'legacy' technologies, that is technologies that are outdated and have little potential to enable electronic filing or 'add on' technologies such as video conferencing. The 2020 global pandemic exposed some of these legacy technology issues as many courts struggled to adapt to enable work to be carried out remotely.[1] Such work was not limited to using technology to create online hearings and events, as it also included the capacity for courts to adopt more sophisticated e-filing mechanisms. Notably, some courts, as discussed below, had already developed, or were in the process of developing, more advanced technological arrangements and such courts were able to adapt more readily to remote working and other arrangements as the global pandemic emerged.[2]

Lawyers can use technological supports quite differently; for example, many lawyers may use sophisticated e-discovery software, project planning software, adapted and adaptive precedent technologies, as well as supportive technologies linked to legal research, billing, and case management. By 2020, a number of publishing firms,[3] as well as private operators,[4] supported such developments in the legal research, guidance, and investigative areas.

By 2020, there were also some significant developments in the online dispute resolution (ODR) area. Many such developments were applied to complaint handling and dispute resolution systems that can exist outside of courts and which may deal with a significant number of disputes.[5] ODR involves the use of digital technology by parties to a dispute and/or a third party to resolve the dispute.[6] As outlined by Legg, it is a broad term encompassing both ADR, which is conducted online, and systems of online courts.[7] More specifically, Sourdin and Liyanage have noted that ODR can include facilitative processes such as online mediation, advisory processes such as online case appraisal, and determinative processes such as online arbitration or adjudication.[8] Sourdin further observes that, in terms of the levels of impact technology has on dispute resolution, it is probably more appropriate to define ODR using the three levels of technologies embedded in this system, including supportive, replacement, and disruptive technologies.[9] Much ODR development has (as with app development more generally), taken place at the first two levels.

However, some ODR can be more disruptive and this is particularly the case where more sophisticated artificial intelligence (AI) is used. ODR can also include processes conducted through a computer program or other AI that do not involve a 'human' practitioner.[10] Such automated processes use coded logic or algorithms to make a decision, part of a decision, or recommendations.[11] As outlined by Parasuraman and Riley, the process of automation is 'characterised by a continuum of levels rather than as an all-or-none concept'.[12] This means decisions can be either wholly or partially automated, with some requiring human involvement at the decision-making stage, and others operating autonomously in lieu of a human decision-maker.[13] Automation can also be integrated at different stages of a decision-making process and involve differing degrees of human oversight and verification.[14]

In terms of the use of ODR by courts, a number of commentators have discussed potential benefits and some courts and tribunals have adopted ODR systems to support court-based dispute management and resolution. Often such developments are linked to automated systems that use basic AI approaches. For example, Justice Perry of the Federal Court of Australia has outlined the 'great benefit' of automated processes: the ability to 'process large amounts of data more quickly, more reliably and less expensively than their human counterparts'.[15]

In China, other developments are focused on how such arrangements can support judicial or court functions so that the emphasis is less on alternative dispute resolution. For example, Justice Yadong Cui, the former President of Shanghai High People's Court, has observed that the value of AI use in China's court system includes assisting judicial handling and improving judicial quality and efficiency, promoting judicial impartiality and credibility, serving public litigation needs, facilitating judicial transparency, enabling the public to 'experience' justice, building a big data analysis platform and improving judges' decision-making.[16]

In general, although the idea of applying AI to legal problems has been investigated since the 1970s, rapid developments in recent years have generated new opportunities.[17]

It seems likely that developments in the ODR area will continue to incorporate more disruptive AI components and will also have more impact in terms of the development of replacement technologies. In this regard, some commentators consider that there is a 50 percent chance of AI outperforming humans in all tasks in 45 years, and of automating all human jobs in 120 years.[18] A 2016 survey of 352 machine learning researchers also found that 45 percent viewed high-level machine intelligence as having a 'good' or 'extremely good' outcome on humanity over the long-run. By contrast, only 10 percent thought it would have a 'bad outcome' and 5 percent an 'extremely bad' outcome.[19] On the other hand, challenges arising from a lack of human oversight of newer technology-embedded processes and decision-making have been noted by Sourdin, Li, and Hinds.[20] Further, in the aforementioned 2016 survey, 48 percent of researchers were of the view that society should prioritise research aimed at minimising the risks of AI.[21]

ODR growth areas

ODR has grown significantly in response to local and international factors in a number of countries. In Australia, for example, Sourdin and Liyanage have noted that this growth is partly attributable to a healthy ADR environment.[22] A needs assessment conducted in Victoria in 2003 found that more than 70 percent of respondents were willing to try ODR to settle a dispute.[23] Given the proliferation of personal media devices and significant increase in technology use over the past decade (see Chapter 1),[24] it is likely that willingness to engage in ODR processes has continued to grow.

In particular, younger people may have a preference for ODR as they are more likely to engage in online activities and transactions. The connection between age and internet connectivity has been substantiated by recent research conducted by the ABS which found that in 2016–2017, persons in the 15–17 years age group were the highest proportion of internet users (at 98 percent uptake), compared to the older persons age group (65 or over), which had the lowest proportion of internet users (at 55 percent uptake).[25]

There are many examples of ODR advisory and determinative processes which go beyond providing information, instead taking a proactive role in facilitating the resolution of disputes.[26] For example, in the energy sector, survey results indicate a 'surprising' support by international energy firms for ODR where advisory and bidding technologies have been used.[27] PayPal and eBay's ODR systems consider roughly 60 million matters per year.[28] In the U.S., commercial ODR operator *Modria* has finalised more than one billion disputes. In Utah's Small Claims Court, an ODR system adopted in September 2018 is capable of handling an entire dispute online. According to Utah Supreme Court Justice Deno Himonas, the introduction of the system is grounded in the Court's commitment to access to justice.[29]

ODR has also recently been embraced on a much larger scale by the European Union (EU). EU Regulation 524/2013 led to the creation an ODR tool to assist consumers and retailers with consumer disputes. In December 2017, a year after the initiation of the service, there were 1.9 million visitors to the service and 24,000 consumer complaints made.[30]

In China, a nationwide effort to establish 'smart' courts and move dispute resolution processes (including litigation) online has seen the wider application of various technologies, including blockchain (in the preservation and authentication of e-evidence), big data (in the collection of real-time trial and enforcement data from all courts in the country), and AI (in the automatic generation and correction of judicial documents).[31]

A number of other international ODR projects also act as 'add-on' systems to traditional justice systems. For instance, the Civil Resolution Tribunal ('CRT'), established in British Columbia in 2012,[32] deals with small claims and condominium disputes, as well as motor vehicle accident and injury claims.[33] The CRT provides tailored legal information, tools, and resources to help parties resolve their dispute. If this fails, the dispute is handed to a facilitator and if an agreement is not reached, the dispute proceeds to adjudication by a tribunal member.[34] As of July 2019, the CRT reported 10,461 completed disputes, with only 1,717 of these disputes resorting to adjudication.[35]

Despite the growth in ODR processes, there remain some significant issues associated with the use of technology, including what has been described as a 'reluctance to innovate' in the justice sector.[36]

Reluctance to innovate

As early as 1999, legal commentators and practitioners had started viewing legal technology as a tool to facilitate an accessible, inexpensive, and efficient legal system.[37] Despite this, Bell has observed that ODR has not 'taken off' to the degree which might be expected considering the pervasive issues of cost and delay, especially in the context of family law litigation.[38] Sourdin and Liyanage have similarly noted that ODR initiatives in the family dispute resolution system remain patchy and are often conducted on a pilot basis.[39]

The literature confirms that there can be a stark contrast between the way courts and private practice have embraced legal technology. On the one hand, some courts are seen as open to technology.[40] For example, Australian Federal Courts, including the Family Court, have set up a Commonwealth Courts Portal (CCP), which allows lawyers to organise and e-file documents, and view files and court decisions.[41] Sourdin and Liyanage have described this as an important development which supports ODR in the family sector.[42] The Australian Federal Courts were some of the first to offer this e-filing service.[43]

China's courts have also demonstrated a strong willingness to apply technologies to assist with various judicial tasks, from generating digital case files to case handling.[44] Writing in relation to his visit to a local court in China's Zhejiang Province in 2017, Susskind reported being 'impressed' with what he saw, including 'a static robot in the reception area that offered online legal help for court users; on-site facilities for the e-filing of documents; dedicated virtual courtrooms; [and] speaker-independent voice recognition ...'.[45]

Despite this, many courts have yet to embrace technologies that can be of assistance and this reluctance can be linked to inadequate budgets and the deprioritising of the justice system in some countries, as well as a lack of innovation readiness.[46] As noted previously, the response to the COVID-19 pandemic by courts around the world highlights the patchy take up of technologies which have resulted in very uneven approaches to court functions in pandemic areas, with many courts effectively closing up operations whilst others have adjusted and flourished.[47] The slow adoption of technology can, as noted above, be attributed to a range of factors and may also relate to judicial conservatism.[48] This reality can mean that many justice apps that are under development or have already been developed are likely to have originated outside of courts and are the result of commercial development (see discussion about the *Adieu* system in Chapter 4).

In the private legal practice area there is also variation and some private practitioners have been viewed as more reluctant to utilise technology, particularly where technologies may support a reduction in cost. Some of this reluctance could be attributed to the pyramid business structure and billable hour arrangements in parts of this sector.[49] This reluctance was also evident in relation to the introduction of ADR arrangements in parts of the justice system in the past. In essence, whilst some practitioners may have embraced change, others have not, particularly where there is a potential 'bottom line' impact. Some private practitioners have, however, embraced technological opportunities and have moved to develop commercial offshoot arrangements, in addition to systems to support clients with AI use in a data driven environment.[50]

According to the Hague Institute for Innovation of Law, one reason for the discontinuation of *Rechtwijzer* – a former ODR tool for separating couples in the Netherlands – was a lack of readiness in the legal profession. Reaching 'a mutually reinforcing partnership with the traditional justice institutions to scale up a platform like *Rechtwijzer*' was described as 'difficult.'[51] Dijksterhuis has summarised some of the reasons for this lack of readiness, including 'the threat of competition, fear of losing work, loss of their familiar way of working and of their autonomy, being replaced by computers in at least part of their work'.[52]

At first glance, this reluctance is surprising, with research showing that technological innovation can provide a range of benefits for lawyers, including improving work efficiency.[53] Sourdin, Li, and Burke, however, have highlighted a 'digital divide' or 'unevenness' in the legal profession in relation to technological innovation.[54] Whilst top tier and large firms with significant budgetary resources are able to invest into incorporating and developing the latest technologies, smaller firms may have more difficulty in doing so.

For example, in 2018, top tier law firm Allens Linklaters launched its 'hub for technology law and innovation', with staff working with university academics 'to explore disruptions to the law, lawyers and the legal system such as reliance on data-driven decision-making and new kinds of biological, artificial and legal "persons"'.[55] By contrast, a 2017 survey of legal practitioners found that smaller firms were less able to invest in AI.[56] Results of this nature give rise to a concern that technological reform will be led by commercial interests or only by the larger law firms and potentially those less concerned with meeting 'justice' objectives.[57]

Conclusions

The development of justice apps and the extent to which such developments will be supported within the justice system is partly dependent on the extent to which judges, courts, and legal practitioners will support such developments. Past developments suggest that the take-up of newer technologies is patchy and developments in ODR may be more likely to take place in respect of complaint and ADR systems that are often located outside courts. In addition, perhaps as in the health area where health apps have proliferated, the ongoing development of justice apps will be driven by commercial interests and end users.

The authors note that even where technology has been embraced by the legal profession, such utilisation has not necessarily been efficient or effective. A 2014 survey conducted by the Family Court and Federal Circuit Court of Australia found that only 30 percent of users agreed, or strongly agreed, that the CCP was 'of assistance'. By contrast, 19 percent 'neither agreed or disagreed', 'disagreed', or 'strongly disagreed'. The portal was not applicable to 61 percent of users.[58] There is also no evidence that the technology being implemented has substantially increased the timeliness of decision-making in the courts.[59] According to Ryan and Evers, reliability and speed of processing systems are the main factors determining the success of technological programs.[60]

The issues surrounding the adoption of newer technologies by courts are complex.[61] Martin, writing in the U.S. context, claims that legal technologies are not analogous to e-commerce developments because the courts are not subject to the same commercial pressures as private enterprise and have entrenched roles and practices.[62] Marilyn Warren, the former Chief Justice of the Supreme Court of Victoria in Australia, has also observed that courts' democratic duties to balance efficiency, save costs, and provide open and impartial justice, mean that not all technologies will be appropriate for the courtroom.[63] There are also concerns expressed by judges in some jurisdictions where apps designed to track judicial preferences and outcomes have been outlawed.[64]

In some countries, however, a more uniform adoption of new technologies by courts has been proposed. For example, the Supreme People's Court (SPC) in China has taken steps to roll out a 'smart court' system across the country by relying on embedded technologies such as big data and AI. However, academics have already identified several risks, including data safety, litigant privacy,[65] and uneven abilities when it comes to adopting technologies, particularly among courts in regions enjoying different levels of financial and other resources.[66]

The 2020 global pandemic has also resulted in a rethink by courts and others of what might be possible in terms of the adoption of newer technologies by courts and legal practitioners. It is likely, for example, that many courts that were already considering the successes of the CRT in Canada will continue to accelerate their adoption of ODR arrangements. Indeed, for many courts and tribunals, steps to integrate ODR systems were already underway prior to 2020. The extent to which such arrangements will incorporate justice apps that may support court users is, however, somewhat more questionable. If the primary approach remains replicating existing court processes, then it seems likely that many apps will only have limited benefits. If, however, the redesign incorporates human centred approaches,[67] and more extensive redesign features, it seems more likely that any apps will support the extension of ODR arrangements, enabling courts and practitioners to operate in different and more effective ways in the coming years.

Notes

1 Tania Sourdin and John Zeleznikow, 'Courts, Mediation and COVID-19', *Australian Business Law Review* 48 (2020) 138.
2 Tania Sourdin and John Zeleznikow, 'Courts, Mediation and COVID-19', *Australian Business Law Review* 28 (2020) 138.
3 See for example Thomson Reuters at <https://legal.thomsonreuters.com/en>.
4 See for example ROSS, the online legal researcher at <https://www.rossintelligence.com/>.
5 This is discussed in more detail in Tania Sourdin, 'A Broader View of Justice' in Michael Legg (ed), *Resolving Civil Disputes* (LexisNexis, 2016).
6 Melissa Conley Tyler and Mark McPherson, 'Online Dispute Resolution and Family Disputes' (2006) 12(2) *Journal of Family Studies* 165, 167.
7 Michael Legg, 'The Future of Dispute Resolution: Online ADR and Online Courts' (2016) 27(4) *Australasian Dispute Resolution Journal* 227, 227.
8 Tania Sourdin and Chinthaka Liyanage, 'The Promise and Reality of Online Dispute Resolution in Australia' in Mohamed S Abdel Wahab, Ethan Katsh and Daniel Rainey (eds), *Online Dispute Resolution: Theory and Practice a Treatise on Technology and Dispute Resolution* (Eleven International Publishing, 2012) 483, 484.
9 Tania Sourdin, *Alternative Dispute Resolution* (Sixth Edition) (Thomson Reuters, 2020), 402.
10 Tania Sourdin and Chinthaka Liyanage, 'The Promise and Reality of Online Dispute Resolution in Australia' in Mohamed S Abdel Wahab, Ethan Katsh and Daniel Rainey (eds), *Online Dispute Resolution: Theory and Practice a Treatise on Technology and Dispute Resolution* (Eleven International Publishing, 2012) 483, 484.
11 Australian Government, *Automated Assistance in Administrative Decision making: Better Practice Guide* (February 2007) 4.
12 Raja Parasuraman and Victor Riley, 'Humans and Automation: Use, Misuse, Disuse, Abuse' (1997) 39(2) *Human Factors* 230, 232.
13 Justice Melissa Perry, 'iDecide: Administrative Decision making in the Digital World' (2017) 91 *Australian Law Journal* 29, 29.
14 Justice Melissa Perry, 'iDecide: Administrative Decision making in the Digital World' (2017) 91 *Australian Law Journal* 29, 29–30.
15 Justice Melissa Perry, 'iDecide: Administrative Decision making in the Digital World' (2017) 91 *Australian Law Journal* 29, 30. See also John Zeleznikow, 'Methods for Incorporating Fairness into the Development of an Online Family Dispute Resolution Environment' (2011) 22(1) *Australasian Journal of Dispute Resolution* 16, 16.

16 Yadong Cui, *"Artificial Intelligence" Makes the Court System More Just, Efficient and Authoritative* <https://law.stanford.edu/china-law-and-policy-association-clpa/articles/>.

17 Felicity Bell, 'Family Law, Access to Justice, and Automation' (2019) 19 *Macquarie Law Journal* 103, 103.

18 Katja Grace et al., 'When Will AI Exceed Human Performance? Evidence from AI Experts' (2018) 62 *Journal of Artificial Intelligence Research* 729, 729.

19 Katja Grace et al., 'When Will AI Exceed Human Performance? Evidence from AI Experts' (2018) 62 *Journal of Artificial Intelligence Research* 729, 733.

20 Tania Sourdin, Bin Li and Tom Hinds, 'Humans and Justice Machines: Emergent Legal Technologies and Justice Apps' (2020) 156 *Precedent* 20.

21 Katja Grace et al., 'When Will AI Exceed Human Performance? Evidence from AI Experts' (2018) 62 *Journal of Artificial Intelligence Research* 729, 733.

22 Tania Sourdin and Chinthaka Liyanage, 'The Promise and Reality of Online Dispute Resolution in Australia' in Mohamed S Abdel Wahab, Ethan Katsh and Daniel Rainey (eds), *Online Dispute Resolution: Theory and Practice a Treatise on Technology and Dispute Resolution* (Eleven International Publishing, 2012) 483, 483.

23 Melissa Conley Tyler, Di Bretherton and Brock Bastian, *Research into Online Alternative Dispute Resolution: Needs Assessment* (Report, 2003).

24 See, eg, Australian Bureau of Statistics, *Household Use of Information Technology, Australia, 2016–17* (Catalogue No 8146.0, 28 March 2018).

25 Australian Bureau of Statistics, *Household Use of Information Technology, Australia, 2016–17* (Catalogue No 8146.0, 28 March 2018).

26 Ayelet Sela, 'Can Computers Be Fair? How Automated and Human-Powered Online Dispute Resolution Affect Procedural Justice in Mediation and Arbitration' (2018) 33 *Ohio State Journal on Dispute Resolution* 91, 100.

27 See Out-Law News, *Survey of International Energy Firms Reveals 'Surprising' Support for Online Dispute Resolution, Says Expert* https://www.pinsentmasons.com/out-law/news/survey-of-international-energy-firms-reveals-surprising-support-for-online-dispute-resolution-says-expert.

28 Tania Sourdin, *Alternative Dispute Resolution* (Lawbook Co, 5th ed., 2016) 393.

29 See Deno Himonas, 'Utah's Online Dispute Resolution Program' (2018) 122(3) *Dickinson Law Review* 875, 881.

30 European Commission, *Report from the Commission to the European Parliament and the Council on the Functioning of the European Online Dispute Resolution Platform Established under Regulation (EU) No 524/2013 on Online Dispute Resolution for Consumer Disputes* (Report, 2017) 4. See also discussion in Chapter 10, Tania Sourdin, *Alternative Dispute Resolution* (Sixth Edition) (Thomson Reuters, Australia, 2020).

31 See Guodong Du, Meng Yu, 'China's Supreme Court Issues a White Paper on Chinese Courts and Internet Judiciary', *China Justice Observer*, https://www.chinajusticeobserver.com/a/supreme-peoples-court-issues-a-white-paper-on-china-court-and-internet-judiciary.

32 *Civil Resolution Tribunal Act*, BC (2012) c 25.

33 See Peter Kenneth Cashman and Eliza Ginnivan, 'Digital Justice: Online Resolution of Minor Civil Disputes and the Use of Digital Technology in Complex Litigation and Class Actions' (2019) 19 *Macquarie Law Journal* 39, 44.

34 Michael Legg, 'The Future of Dispute Resolution: Online ADR and Online Courts' (2016) 27(4) *Australasian Dispute Resolution Journal* 227, 230.

35 Civil Resolution Tribunal, *CRT Statistics Snapshot – July 2019* (Web Page, 6 August 2019) <https://civilresolutionbc.ca/crt-statistics-snapshot-july-2019/>.

36 See Tania Sourdin, Bin Li and Tony Burke, 'Just Quick and Cheap? Civil Dispute Resolution and Technology' (2019) 19 *Macquarie Law Journal* 17, 32.

37 Law Reform Committee, Parliament of Victoria, *Technology and the Law* (May 1999) [3.1].

38 Felicity Bell, 'Family Law, Access to Justice, and Automation' (2019) 19 *Macquarie Law Journal* 103, 120.

39 Tania Sourdin and Chinthaka Liyanage, 'The Promise and Reality of Online Dispute Resolution in Australia' in Mohamed S Abdel Wahab, Ethan Katsh and Daniel Rainey (eds), *Online Dispute Resolution: Theory and Practice a Treatise on Technology and Dispute Resolution* (Eleven International Publishing, 2012) 483, 499.

40 Frederika De Wilde, 'Courtroom Technology in Australian Courts: An Exploration into its Availability, Use and Acceptance' (2006) 26 *Queensland Lawyer* 303, 304.

41 Family Court of Australia, *Commonwealth Courts Portal* (Web Page, 2019) <http://www.familycourt.gov.au/wps/wcm/connect/fcoaweb/how-do-i/ccp/register-for-ccp/>.

42 Tania Sourdin and Chinthaka Liyanage, 'The Promise and Reality of Online Dispute Resolution in Australia' in Mohamed S Abdel Wahab, Ethan Katsh and Daniel Rainey (eds), *Online Dispute Resolution: Theory and Practice a Treatise on Technology and Dispute Resolution* (Eleven International Publishing, 2012) 483, 498.

43 Philippa Ryan and Maxine Evers, 'Exploring eCourt innovations in New South Wales Civil Courts' (2016) 5 *Journal of Civil Litigation and Practice* 65, 66.

44 Supreme People's Court of China, *Chinese Courts and Internet Judiciary (2019)*, 79–83.

45 Richard Susskind, *Online Courts and The Future of Justice* (2020), Oxford University Press, 170–1.

46 See Tania Sourdin, Bin Li and Tony Burke, 'Just Quick and Cheap? Civil Dispute Resolution and Technology' (2019) 19 *Macquarie Law Journal* 17, 32.

47 Tania Sourdin and John Zeleznikow, 'Courts, Mediation and COVID-19', *Australian Business Law Review* 48 (2020) 138.

48 See Tania Sourdin, *Judges, Technology and AI*, forthcoming, (2021) (Edward Elgar).

49 Tania Sourdin, Bin Li and Tony Burke, 'Just Quick and Cheap? Civil Dispute Resolution and Technology' (2019) 19 *Macquarie Law Journal* 17, 32.

50 See for example, Allens Linklaters 'AI Toolkit – Ethical, Safe, Lawful' (2019) available at <https://www.allens.com.au/globalassets/pdfs/campaigns/report-ai-toolkit_may19.pdf>.

51 Bregje Dijksterhuis, 'The online divorce resolution tool Rechtwijzer uit Elkaar examined' in Mavis Maclean and Bregje Dijksterhuis (eds), *Digital Family Justice: From Alternative Dispute Resolution to Online Dispute Resolution?* (2019), Hart Publishing, 205–10.

52 Bregje Dijksterhuis, 'The Online Divorce Resolution Tool Rechtwijzer uit Elkaar Examined' in Mavis Maclean and Bregje Dijksterhuis (eds), *Digital Family Justice: From Alternative Dispute Resolution to Online Dispute Resolution?* (2019), Hart Publishing, 206.

53 See, eg, John Zeleznikow, 'Don't Fear Robo-Justice. Algorithms Could Help More People Access Legal Advice', *The Conversation* (23 October 2017) <http://theconversation.com/dont-fear-robo-justice-algorithms-could-help-more-peopleaccess-legal-advice-85395>.

54 Tania Sourdin, Bin Li and Tony Burke, 'Just Quick and Cheap? Civil Dispute Resolution and Technology' (2019) 19 *Macquarie Law Journal* 17, 32.

55 Anna Collyer and Ian McGill, *Allens Hub for Technology, Law & Innovation Launches to Confront the Future of Law* (Web Page, 24 November 2017) <https://www.allens.com.au/med/pressreleases/ pr24nov17.htm>.

56 Macquarie Bank, *An Industry in Transition 2017: Legal Benchmarking Results* (Report, 2017) <https://www.macquarie.com/dafiles/Internet/mgl/au/docs/noindex/macquarie-2017-legal-benchmarking-full-results.pdf> 28.

57 Tania Sourdin, Bin Li and Tony Burke, 'Just Quick and Cheap? Civil Dispute Resolution and Technology' (2019) 19 *Macquarie Law Journal* 17, 31.

58 Family Court of Australia and Federal Circuit Court of Australia, *Court User Satisfaction Survey* (Report, 2015).

59 Justice Stuart Morris, 'Where Is Technology Taking the Courts and Tribunals?' (2005) 15 *Journal of Judicial Administration* 17, 20.

60 Philippa Ryan and Maxine Evers, 'Exploring eCourt Innovations in New South Wales Civil Courts' (2016) 5 *Journal of Civil Litigation and Practice* 65, 68.

61 See Tania Sourdin, *Judges, Technology and AI*, forthcoming (2021) (Edward Elgar).

62 Peter W Martin, 'How Structural Features of the U.S. Judicial System Have Affected the Take-Up of Digital Technology by Courts' (2010) 1(1) *European Journal of Law and Technology* [3].

63 Justice Marilyn Warren AC, 'Embracing Technology: The Way Forward for the Courts' (2015) 24 *Journal of Judicial Administration* 227, 229.

64 See Tania Sourdin, *Judges, Technology and AI*, forthcoming (2021) (Edward Elgar) referring to developments in France.

65 Jiao Feng and Ming Hu, 'Smart Justice: A New Pathway to Justice and Its Limits', *Zhejiang Social Sciences* (2018), Issue 6, 72–3.

66 Tao Wu, Man Chen, The Construction of Smart Courts: Values and Framework Design, *Social Sciences* (2019), Issue 5, 106–7.

67 Lisa Toohey et al, 'Meeting the Access to Civil Justice Challenge: Digital Inclusion, Algorithmic Justice, and Human-Centred Design' (2019) 19 *Macquarie Law Journal* 133, 145.

3 Justice apps – objectives and opportunities

Introduction

Enhancing access to justice is a primary objective of many online dispute resolution (ODR) processes and justice apps, although some are of course also focussed on returning a profit to their (mainly) commercial developers. Although strategies to address access to justice have historically focussed on modifying court processes and improving access to legal representation, researchers have recognised that access to justice requires more than just access to courts and lawyers (see also discussion in Chapter 2). A number of researchers have explored the access to justice benefits associated with incorporating technology into dispute resolution, with substantial claims made about the capacity of artificial intelligence (AI) or automated systems to improve access to justice.[1]

Internationally, it has also been recognised that technology can create new pathways to justice,[2] with the former Chief Justice of the Supreme Court of Canada, Beverley McLachlin, urging the legal profession to accept the reality that some tasks traditionally performed by lawyers can now be more effectively executed through technological means.[3]

First and foremost, improved access to justice can come in the form of technology reducing cost and delay.[4] Bellucci, Macfarlane, and Zeleznikow have highlighted the affordability of online technologies compared to the costs of litigation or prolonged ADR.[5] More specifically, Sourdin, Li, and Burke have observed that ODR can save travel time and disbursements, alongside contributing to a faster finalisation of disputes compared with traditional litigation processes and traditional forms of ADR.[6] Justice Perry of the Federal Court of Australia has also noted that automated systems can assist self-represented applicants that come before the courts and tribunals in accessing justice.[7]

Cost savings can result from clients being able to do some or all of their legal work themselves ('unbundling' – see discussion in Chapter 6), or through lawyers passing cost savings on to their clients as technology assists them to work more efficiently.[8] According to McGill, Bouclin, and Salyzyn, however, apps designed for the general public, as opposed to lawyer-oriented apps, 'hold more promise for improving access to justice'.[9] This chapter explores barriers to access to justice in the context of apps and the extent to which developments in this field can support improved access.

Cost, delay, and access

Clearly cost issues can have a significant impact on the extent to which people can resolve or finalise disputes in the justice sector. These issues can be less significant in some jurisdictions; for example, in inquisitorial court systems there may be less focus on the costs generated than in adversarial systems (which will often require a greater focus on advocacy costs).[10] In the civil dispute area (i.e., civil and commercial disputes) many studies have indicated that cost is a major factor in determining whether people are able to defend their rights or commence an action.[11]

In the family law context, cost issues can be a particularly significant barrier for clients. As noted by McGill, Bouclin, and Salyzyn, it is not uncommon for individuals experiencing a divorce to find that retaining a lawyer is beyond their financial means.[12] Data shows that complex family law cases in Australia can cost parties more than $200,000.[13] The Australian Institute of Family Studies has reported that 71 percent of divorcing and separating couples in Australia currently finalise financial settlements without access to legal advice.[14] Further, it has been recognised that the Australian family law system is plagued by delays and backlogs.[15] As stated by Bell, in the family law context, delays can have a severe impact, not just on the parties, but also on their children.[16] Beyond Australia, many other jurisdictions have raised concerns about both cost and delay.[17]

In March 2019, the Australian Law Reform Commission (ALRC) released its final report in its inquiry into Australia's family law system. The inquiry identified serious structural and systemic difficulties within the current system. Whilst highlighting cost and delay, the ALRC also indicated that navigating the justice system was difficult for clients. Structurally, it highlighted the problems with the existing bifurcated system in Australia which requires parties to navigate multiple courts across jurisdictions.[18]

At a systemic level, the ALRC found the current family law system in Australia suffers from two major deficiencies: (i) 'impenetrable' legislation which is not sufficiently transparent or accessible, hindering the ability of parties to understand 'the parameters within which decisions will be made about their children and their property and financial interests'; and (ii) a lack of resources at all stages of the process.[19] Improving dispute resolution processes, including through the early, cost effective, and less adversarial resolution of parenting and property disputes, was one focus of the ALRC's inquiry.[20] In addition, the ALRC considered how forms were used and how guidance could be provided to disputants. The authors note that each of these targeted areas of improvement can be supported by apps (see also Chapter 4).

Fairness and access

Whilst perceptions of justice can be linked to cost savings, there is also some concern that a focus on cost reduction and time saving can result in the system becoming less 'just', especially where justice processes are 'dehumanised'.[21] This concern is discussed in greater detail below (see Chapter 5). Some academics have also queried whether automation should be viewed as a panacea for access to justice issues.[22] As noted by Bell, there are many reasons why people do not access justice options, beyond just affordability. These include 'not knowing there is a legal issue, personal stress or distress, inconvenience, fear or mistrust of the legal system, or lacking faith in the system's effectiveness'. Nevertheless, it is acknowledged that affordability plays a key role in determining whether people can access the justice system.[23]

Further, McGill, Bouclin, and Salyzyn have identified two other opportunities presented by justice apps beyond the mitigation of financial barriers: (i) the mitigation of psychological and informational barriers; and (ii) the mitigation of physical barriers.[24] They also noted that in light of these opportunities, justice apps 'hold promise for contributing to an increase in client empowerment'.[25]

In terms of mitigating psychological and informational barriers, Gershowitz has referred to justice apps as a 'democratising force' due to their ability to inform individuals and assist them in exercising their rights.[26] Other researchers have similarly highlighted the potential for technology to 'demystify' legal institutions.[27] McGill, Bouclin, and Salyzyn have further noted that justice apps can provide 'more holistic or client-centered assistance' than that offered by lawyers giving solely legal advice, thus 'filling a set of law-related needs not currently provided by the conventional legal services market'.[28]

Finally, when it comes to mitigating informational barriers, Sourdin has noted that apps are an especially useful tool for communicating to young people who commonly use their smartphones to access information.[29] Looking at legal aid clients, Dysart has also observed that while technology and courts may intimidate clients, they are often well versed in the use of smartphones.[30] A similar benefit is identified on the website of the AI chatbot *LISA*, which states that the app can provide a solution for those who do not wish to use a human lawyer because they are intimidated by the use of legalese, or fear feeling intellectually inferior.[31]

The mitigation of physical barriers has also been identified by a number of commentators as a key benefit of justice apps.[32] The fact that disputes can include international, national, and local interaction, means that ODR is capable of exerting a significant influence on the justice system, particularly in countries with a relatively sparse population, such as Australia.[33] McGill, Bouclin, and Salyzyn have noted that 'apps can be accessed on an anytime, anywhere basis'. This flexibility is seen as particularly important for those living in remote areas who may struggle to obtain appropriate in-person legal information and services.[34]

Improved access to justice can also remove barriers that prevent disadvantaged parties from accessing dispute resolution processes.[35] When it comes to supportive technologies, for instance, improved access to justice flows from the 'weaker' party in the dispute being able to receive appropriate legal information and advice, thereby addressing power imbalances and increasing the possibility of obtaining a 'just' settlement.[36] More broadly, it has been recognised that ODR can provide a favourable alterative to those disadvantaged by the fact that 'traditional dispute resolution mechanisms advantage people who are physically attractive, articulate, well-educated, or members of a dominant ethnic, racial, or gender group'.[37]

Rogers has similarly noted that ODR 'presents promising possibilities for reaffirming victim autonomy, increasing victim safety, and reducing the effect of harmful gender and racial norms in the judicial process'.[38] An evaluation of *Rechtwijzer* in the Netherlands, also revealed positive results for user empowerment. It was reported that 84 percent of surveyed participants who used the platform felt they had increased control over their separation, including as a result of being able to 'spread out' the hours they spent dealing with each step of the divorce process. Further, over half the participants reported experiencing 'low' or 'very low' stress levels during their separation.[39]

In general, there has been less consideration of the possibilities that justice apps, as opposed to ODR or technology more broadly, can offer in terms of improving access to justice. However, there is some relevant material. For example, in the U.S., the key objective of the *Apps for Justice Project* – which sought to improve the accessibility of legal services in the state of Maine – was to increase access to justice by creating apps that allowed low and moderate-income consumers to address legal issues, both independent of, and with professional assistance, and to assist legal firms in handling a larger volume of low-income clients.[40]

The *Apps for Justice Project* demonstrates that reducing the cost of accessing legal help, and by extension, improving access to justice, can be a key design objective of justice apps. Another U.S. app, *Gideon*, which aims to connect clients and lawyers, is stated to be designed 'specifically for the access to justice community', helping legal aid providers 'determine who should get what help, when, how, and by whom to maximize client outcomes per dollar spent'.[41] The website for the AI chatbot, *LISA*, similarly states that the app's aim is to 'take a bite out of the access to legal services and access to justice crisis'.[42]

Some justice apps that are oriented towards access to justice are supportive in that their focus is on alerting people to developments in the legal sector and enabling a greater general understanding of the legal sector at low or even no cost. For example, in 2015, China's SPC launched the free of charge *China Court Mobile TV* (*Zhong Guo Fa Yuan Shou Ji Dian Shi*) app, with the aim of promoting open justice and disseminating useful information to the public.[43] This app has five areas of focus: legal news, hot topics, live trials, press conferences, and judge talks. 'Legal News' reports on laws and the important work of courts across the country, while 'Hot Topics' provides in-depth follow-up and analysis on high-profile cases in China. 'Live Trials' and 'Press Conferences' enable app users to access certain open court trials, and SPC and local court briefings, respectively. In 'Judge Talks', an online classroom model is adopted where selected judges across the country educate the general public through discussion of legal issues. Chief Justice Qiang Zhou, President of SPC, commented during the app launch ceremony that the app would enable the general public to 'better understand justice, participate in justice and supervise justice'.[44]

Despite the opportunities for improved access to justice offered by apps, there have been some concerns raised that justice apps may skew the access to justice debate. As outlined by McGill, Bouclin, and Salyzyn, justice apps which are marketed as cheaper and easier substitutes for full-service legal representation may reduce the general sense of urgency about the access to justice crisis and distract from the on-going need to improve the affordability and accessibility of real-time legal and court services.[45] However, such concerns may relate more to specific apps, rather than apps that provide more generalised support and information (such as noted above in relation to China).

Functions

In terms of generalised information apps, in countries such as Australia, justice apps have primarily been used to provide general information on a topic, as opposed to specific advice on an individual case. For instance, the *LegalAidSA* app, discussed above, provides general information on areas that the clients of Legal Aid South Australia commonly seek advice (e.g., family law, traffic offences, wills).[46]

Internationally, a Canadian taxonomy of 60 justice apps found that they perform a variety of functions, including: providing legal information and advice, creating documents, streamlining conventional legal processes, and collecting evidence.[47] It was also reported that justice apps target two primary users: (i) lawyers; and (ii) the general public.[48] In relation to lawyers, justice apps can promote more efficient legal service delivery and assist to streamline legal research.[49] For instance, *Rangefindr* is a Canadian tool designed to assist lawyers and judges in determining appropriate sentencing ranges in criminal matters.[50] Tools targeted at lawyers may also use machine learning to provide answers to legal questions. An example is *Blue J Legal*, based in New York, which analyses fact situations using deep learning algorithms to provide 'authoritative answers' on specific legal subjects.[51]

In terms of apps targeted at public users, a Canadian study found that these fall within three broad categories:

(i) apps designed to make legal services easier to access (e.g., by connecting individuals with lawyers);
(ii) apps seeking to materially change the way individuals interact with the legal system, usually by allowing users to engage with self-help processes;
(iii) apps seeking to provide legal self-help tools and assistance to the public that may not typically be offered by a lawyer.[52]

The second category – apps which change the way individuals interact with the legal system – can be further divided into four sub-categories: (i) apps that offer general legal information on a specific subject; (ii) apps that allow users to create legal documents; (iii) apps that streamline conventional legal processes; and (iv) apps that help individuals with legal research.[53] In the family law context, an example of the third sub-category – apps that streamline conventional legal processes – is *Thistoo*, a 'personal divorce assistant' which previously operated in Ontario, Canada, and walked users through the necessary steps of an uncontested divorce.[54]

Jurisdictional responses vary in terms of app development. For example, in China many developments are focussed on the first category noted above and such apps tend to be developed by courts. This is partly because the SPC has taken the lead in building 'smart courts' throughout the country by introducing newer technologies into the justice sector since 2016. Local courts at various levels have responded to this initiative by developing their own online platforms and apps to enable judges, the general public and lawyers to engage with each other.[55] For the public users, some apps are for educational purposes and provide information in regard to legislation (i.e., legal database apps). For example, the *China Court Mobile TV* app, discussed above, is an educational tool which provides the public with legal knowledge.

Another example in China is a free app called *Compilation of Chinese Laws* (*Zhong Guo Fa Lv Hui Bian*) which offers users more than 1,000 Chinese laws that are of relevance to daily life, including the Constitution of China, contract law, and marriage law. However, legal professionals usually pay for more sophisticated apps, such as *Read Law* (*Kan Fa*), to access regularly updated legislation information.

More recently in China, there have been some developments in the second and third category areas noted above. In this regard, an app called *Ning Bo Mobile Mini Court* (*Ning Bo Yi Dong Wei Fa Yuan*) was officially launched by Ning Bo Intermediate People's Court in Zhejiang Province in January 2018 and enables litigants to complete the whole litigation process, including case filing, serving legal documents, mediation,[56] evidence exchange, court hearings, and any follow up enforcement. As of August 2018, approximately 70,000 cases had been filed using this app and it was reported that this tool had saved judicial costs and enhanced litigant satisfaction.[57] Because of the success of this app in the Ning Bo region, the SPC continued to develop a national version of *Mobile Mini Court* (as opposed to the regional version in Ning Bo) and promoted the new version to other parts of the country from August 2018. In January 2020, Chief Justice Zhanguo Li, President of Zhejiang High People's Court, observed that *Mobile Mini Court* in Zhejiang Province had already dealt with over 1.36 million cases involving around 470,000 litigants and about 90,000 lawyers.[58]

Justice apps have also been studied in the context of the criminal justice sector. A study carried out in the U.S. used a somewhat similar taxonomy and found that criminal justice apps fall within four broad categories: (i) apps which teach individuals about the law;

(ii) apps which help suspects and defendants connect with lawyers; (iii) apps which help defendants navigate the court system; and (iv) 'reform apps' or 'paradigm-shifting apps' which seek to bring systemic changes to the criminal justice system.[59] Interestingly, the study concluded that criminal-justice apps are unlikely to progress to the extent of fully assisting people to navigate the criminal justice system.[60] This is because each case is different and 'nuanced analysis is often critical'. Further, 'lawyers practice for years to become experts in all the procedural steps and motions' and there are often 'procedural variations from judge to judge'. All of this means that arguably 'criminal law and procedure cannot be simplified in a cell phone application, except at a very high level of generality'.[61]

The evaluation of apps and a client focus

There is a dearth of research evaluating justice apps or developing appropriate frameworks to assess their utility. Nevertheless, it is likely that frameworks developed to evaluate health apps can be to some extent adapted to the justice app context.[62] The American Psychiatric Association (APA) has identified a number of reasons for the evaluation and rating of mental health apps, including the risk of apps offering incorrect or misleading information, failing to properly protect user information, selling user data without appropriate user consent, or collecting more information than is necessary for the app to fulfil its functions.[63]

In the context of mental health apps, the APA has also developed an app evaluation and rating model.[64] Apps are assessed on the basis of: (i) privacy and security considerations; (ii) effectiveness; (iii) ease of use; and (iv) interoperability. Based on these considerations, users are advised to either not proceed with the app, proceed with caution, or proceed freely. In terms of privacy and security, relevant considerations relate to the existence of a privacy policy, the type of data that is collected, and whether the data is de-identified. The privacy and security concerns associated with justice apps are considered in more detail in Chapter 5 of this book.

In terms of app effectiveness, the APA's model gives consideration to user feedback, evidence of potential benefits, whether the app does what it claims to do, and whether the content is of reasonable value. The third criterion – ease of use – requires consideration of the app's interface and overall functionality. Outside the mental health context, an empirical study of mobile banking apps has similarly made clear that visual design, navigation simplicity, perceived usefulness, and performance are important to app users.[65]

The New Zealand Ministry of Health has also created an app evaluation guide, assessing app quality on the basis of: (i) engagement (including customisability, interactivity, and suitability for the target audience); (ii) functionality (including performance, ease of use, and navigation); (iii) aesthetics; and (iv) information quality.[66] In regard to the latter, Kapoo and Vij have reported that information design has the strongest influence on customer ratings of apps. This covers the accuracy, currency, and comprehensiveness of information.[67]

The above app evaluation frameworks can readily be adapted to the justice app context (see the conclusions in this chapter). An evaluation of the *PENDA* app during its testing stage found that accessibility and user-friendliness could be improved with greater use of plain English, a search function, filtering/customisation options, the use of icons and pictures to break up text, page translation options to increase potential user groups, and the more prominent placing of resources and referrals (see also Chapter 5).[68]

More specifically in the legal context, Cruz has highlighted the need for 'culturally competent' legal technology, noting that 'legal technological innovations can help narrow the access to justice gap by using culturally competent design features that make programme applications more accessible to diverse end users'.[69] Cruz explains that culturally competent design principles fall within two broad categories: (i) user experience principles (i.e., 'factors that affect the end user's ability to engage meaningfully with technology, such as ease of use, ability to understand, and value of use of the program'); and (ii) principles related to the mechanics of the technology design (i.e., factors such as the 'coding of AI formulas, hardware and software considerations, and web page or application drafting techniques'). It is noted that without careful coding considerations, legal technologies that integrate AI into their decision-making programmes run the risk of producing biased results.[70] The risk of bias in automated decision-making is considered in more detail in Chapter 5 of this book.

In terms of ODR, Abedi, Zeleznikow, and Brien have considered the development of regulatory standards for security in ODR systems.[71] Consideration of the extent to which justice apps comply with such standards can provide another means of evaluating their overall utility. Using data from qualitative interviews and online surveys with ODR providers and consumers, the authors identify three facets of security in ODR systems: (i) information security (ensuring information is kept secure and confidential); (ii) privacy (personal data protection); and (iii) authentication (ensuring that parties in online transactions or communications are who they claim to be). In line with these three facets, the authors recommend

ODR systems be supported by privacy ethical guidelines and security guidelines created by professional data protection agencies.[72]

It is likely, however, that 'the existence of multiple actors in the ODR world' means there is a need for standards which address the separate activities of these various actors. Specifically, Rainey has noted:

> Standards for practitioners must tell third parties how to behave in relationships with primary parties. Standards for fourth parties (apps and platforms) must set parameters that describe how algorithms and AI must behave in relationships with primary parties. Standards for developers must address a number of issues, including the basic ADR/ODR knowledge that developers should have in addition to their coding skills. Standards for service providers must address issues of privacy and confidentiality in an online world where privacy is, some would argue, an illusion. The fractured nature of the actors and the standards they need will continue to be a challenge.[73]

In terms of privacy and confidentiality (which is discussed in greater detail in Chapter 5) there remain significant issues with app use more generally as both Apple and Google impact on the operation of almost all smartphones. There are various concerns about how data is used, what user data is obtained and to what extent data is passed on to others. As many researchers have found, privacy and data security issues can vary significantly in terms of apps more generally.[74] Whilst concerns in this area can be linked to app permissions, significant variations can also arise as a result of country or regional specific data and privacy protection arrangements.

Conclusions

As noted above, apps can be categorised according to the nature of what they do as well as the audience that engages with justice apps. Importantly, regardless of the nature of an app, having a clear approach in terms of evaluation will support further developments in this area. The authors suggest that such an approach based on the matters noted above could involve considering four factors (each with a number of variables that could be more or less relevant depending on the app characteristics):

(i) **Ease of use**: to what extent are users involved in the design of the app and to what extent does the app support access to justice? Ease of use incorporates the criteria noted previously in relation to health app reviews, that is matters relating to: (i) engagement (including customisability, interactivity, and suitability for the target audience); (ii) functionality (including performance, ease of use, and navigation); and (iii) aesthetics;

(ii) **Effectiveness**: In the justice sector, effectiveness can vary according to the nature of the app being evaluated. For example, an app that is focussed on providing supportive technologies via an information pathway may be evaluated quite differently from an app that has objectives more specifically focussed on the triage of disputes or the provision of expert advice. In general, however, effectiveness includes a number of elements which may be relevant and can include the extent to which the app resolved or limited the dispute, was perceived to be fair, and achieved outcomes that are broadly consistent with public and party interests.[75] Effectiveness in the justice app area also incorporates the notion that the app promotes justice, that is, it supports the dignified treatment of people engaged in the justice system and ensures that human review is available and supported so that substantive justice needs may be met (see discussion in Chapter 5);

(iii) **Privacy and security considerations**: Privacy can relate to how data is stored as well as other factors that are linked to security and other issues (e.g., app permissions and third party data sharing). Again, the extent to which the factors are relevant may vary according to the app focus, the developer interests and the domestic jurisdictional factors that may impact on app arrangements. In basic terms, however, such considerations include ensuring that information is kept secure and confidential, that personal data protections are in place and that authentication arrangements support the app use;

(iv) **Interoperability**: This consideration is linked not only to ensuring that an app may work well and on a range of devices with a range of software supports but also incorporates the notion that the app functions holistically and can be linked effectively to other systems such as court systems where appropriate.

Notes

1 See, eg, Lisa Toohey et al., 'Meeting the Access to Civil Justice Challenge: Digital Inclusion, Algorithmic Justice, and Human-Centred Design' (2019) 19 *Macquarie Law Journal* 133, 143; Tania Sourdin, Bin Li and Tony Burke, 'Just Quick and Cheap? Civil Dispute Resolution and Technology' (2019) 19 *Macquarie Law Journal* 17, 18; Melissa Conley Tyler and Mark McPherson, 'Online Dispute Resolution and Family Disputes'

(2006) 12(2) *Journal of Family Studies* 165. The authors recognize that there are distinct differences in terms of how access to justice is measured see <https://ncforaj.org/>.

2 See, eg, David Luban, 'Optimism, Skepticism and Access to Justice' (2016) 3(3) *Texas A&M Law Review* 495, 502; Jessica Frank, 'A2J Author, Legal Aid Organizations, and Courts: Bridging the Civil Justice Gap Using Document Assembly' (2017) 39(2) *Western New England Law Review* 251; Sherley Cruz, 'Coding for Cultural Competency: Expanding Access to Justice with Technology' (2019) 86 *Tennessee Law Review* 347, 348.

3 Chief Justice Beverley McLachlin, 'The Legal Profession in the 21st Century' (Canadian Bar Association Plenary, Calgary, 14 August 2015).

4 Tania Sourdin, Bin Li and Tony Burke, 'Just Quick and Cheap? Civil Dispute Resolution and Technology' (2019) 19 *Macquarie Law Journal* 17, 17.

5 Emilia Bellucci, Deborah Macfarlane and John Zeleznikow, 'How Information Technology Can Support Family Law and Mediation' in Witold Abramowicz, Robert Tolksdorf and Krzysztof Węcel (eds), *Business Information Systems Workshops* (Springer International Publishing, 2010) 243, 252.

6 Tania Sourdin, Bin Li and Tony Burke, 'Just Quick and Cheap? Civil Dispute Resolution and Technology' (2019) 19 *Macquarie Law Journal* 17, 26.

7 Justice Melissa Perry, 'iDecide: Administrative Decision making in the Digital World' (2017) 91 *Australian Law Journal* 29, 34. See also Jena McGill, Suzanne Bouclin and Amy Salyzyn, 'Mobile and Web-based Legal Apps: Opportunities, Risks and Information Gaps' (2017) 15 *Canadian Journal of Law and Technology* 229, 230.

8 Tania Sourdin, 'Judge v Robot: Artificial Intelligence and Judicial Decision making' (2018) 41(4) *University of New South Wales Law Journal* 1114, 1117–8.

9 Jena McGill, Suzanne Bouclin and Amy Salyzyn, 'Mobile and Web-based Legal Apps: Opportunities, Risks and Information Gaps' (2017) 15 *Canadian Journal of Law and Technology* 229, 240–1.

10 For example, China has the inquisitorial court system. When mentioning the purpose of *Civil Procedure Law of China*, Article 2 of the legislation indicates the 'quick' and 'just' resolution of disputes with no reference to the 'cheap' aspect. The legislation is available at <http://cicc.court.gov.cn/html/1/219/199/200/644.html>. In general, most inquisitorial systems assume that some investigation will be undertaken by a court or the State, rather than the parties to a dispute, thus resulting in increased public cost with reduced party cost.

11 See for example Legal Services Corporation (2017) *The Justice Gap: Measuring the Unmet Civil Legal Needs of Low-income Americans*. Prepared by NORC at the University of Chicago for Legal Services Corporation (Washington, DC) Available at <https://www.lsc.gov/sites/default/files/images/TheJusticeGap-FullReport.pdf>.

12 Jena McGill, Suzanne Bouclin and Amy Salyzyn, 'Mobile and Web-based Legal Apps: Opportunities, Risks and Information Gaps' (2017) 15 *Canadian Journal of Law and Technology* 229, 231–2.

13 Ross Gittins, 'Defeated by High Levels of Legal Costs: The Terrible Injustice Most of Us Could Face', *The Sydney Morning Herald* (29 August 2017) <https://www.smh.com.au/opinion/

defeated-by-high-legal-costs-the-terrible-injustice-most-of-us-could-face-20170829-gy68pr.html>.

14 Lixia Qu et al., 'Post-Separation Parenting, Property and Relationship Dynamics after Five Years' (Report, Australian Institute of Family Studies, December 2014) 115.

15 See, eg, Felicity Bell, 'Family Law, Access to Justice, and Automation' (2019) 19 *Macquarie Law Journal* 103, 113; Chief Justice Alastair Nicholson, 'Legal Aid and a Fair Family Law System' (Legal Aid Forum towards 2010, Australian Capital Territory, 21 April 1999).

16 Felicity Bell, 'Family Law, Access to Justice, and Automation' (2019) 19 *Macquarie Law Journal* 103, 114.

17 See for example Anne Barlow, Rosemary Hunter, Janet Smithson and Jan Ewing: *Mapping Paths to Family Justice: Resolving Family Disputes in Neoliberal Times* (2017) (Palgrave Macmillan).

18 Australian Law Reform Commission, *Family Law for the Future – An Inquiry into the Family Law System: Final Report*, Report No 135 (March 2019) 31–2 [1.7].

19 Australian Law Reform Commission, *Family Law for the Future – An Inquiry into the Family Law System: Final Report*, Report No 135 (March 2019) 32 [1.8].

20 Australian Law Reform Commission, *Family Law for the Future – An Inquiry into the Family Law System: Final Report*, Report No 135 (March 2019) 30 [1.3].

21 Tania Sourdin, Bin Li and Tony Burke, 'Just Quick and Cheap? Civil Dispute Resolution and Technology' (2019) 19 *Macquarie Law Journal* 17, 18.

22 See, eg, Francesca Bartlett, 'An Uncomfortable Place for Technology in the Australian Community Legal Sector' (International Legal Ethics Conference VIII, University of Melbourne Law School, 7 December 2018).

23 Felicity Bell, 'Family Law, Access to Justice, and Automation' (2019) 19 *Macquarie Law Journal* 103, 125.

24 Jena McGill, Suzanne Bouclin and Amy Salyzyn, 'Mobile and Web-based Legal Apps: Opportunities, Risks and Information Gaps' (2017) 15 *Canadian Journal of Law and Technology* 229, 241–3.

25 Jena McGill, Suzanne Bouclin and Amy Salyzyn, 'Mobile and Web-based Legal Apps: Opportunities, Risks and Information Gaps' (2017) 15 *Canadian Journal of Law and Technology* 229, 242.

26 Adam M Gershowitz, 'Criminal Justice Apps' (2019) 5 *Virginia Law Review* 1, 12.

27 See, eg, Jane Bailey, Jacquelyn Burkell and Graham Reynolds, 'Access to Justice for All: Towards an 'Expansive Vision' of Justice and Technology' (2013) 31 *Windsor YB Access Just* 181, 195.

28 Jena McGill, Suzanne Bouclin and Amy Salyzyn, 'Mobile and Web-based Legal Apps: Opportunities, Risks and Information Gaps' (2017) 15 *Canadian Journal of Law and Technology* 229, 242.

29 Tania Sourdin, 'Justice and Technological Innovation' (2015) 25 *Journal of Judicial Administration* 96, 97.

30 Joe Dysart, 'Justice in Your Palm' (2014) 101 *ABA Journal* 54, 55.

31 Robot Lawyer Lisa, *LISA* (2020) <https://robotlawyerlisa.com/>.

32 See, eg, Stefan Lancy, 'ADR and Technology' (2016) 27 *Australasian Dispute Resolution Journal* 168, 171; Tania Sourdin and Chinthaka Liyanage,

'The Promise and Reality of Online Dispute Resolution in Australia' in Mohamed S Abdel Wahab, Ethan Katsh and Daniel Rainey (eds), *Online Dispute Resolution: Theory and Practice a Treatise on Technology and Dispute Resolution* (Eleven International Publishing, 2012) 483, 499.

33 Tania Sourdin, Bin Li and Tony Burke, 'Just Quick and Cheap? Civil Dispute Resolution and Technology' (2019) 19 *Macquarie Law Journal* 17, 36.

34 Jena McGill, Suzanne Bouclin and Amy Salyzyn, 'Mobile and Web-based Legal Apps: Opportunities, Risks and Information Gaps' (2017) 15 *Canadian Journal of Law and Technology* 229, 243.

35 Michael Legg, 'The Future of Dispute Resolution: Online ADR and Online Courts' (2016) 27(4) *Australasian Dispute Resolution Journal* 227, 227.

36 Tania Sourdin, Bin Li and Tony Burke, 'Just Quick and Cheap? Civil Dispute Resolution and Technology' (2019) 19 *Macquarie Law Journal* 17, 27.

37 Melissa Conley Tyler and Mark McPherson, 'Online Dispute Resolution and Family Disputes' (2006) 12(2) *Journal of Family Studies* 165, 169.

38 Sarah Rogers, 'Online Dispute Resolution: An Option for Mediation in the Midst of Gendered Violence' (2009) 24(2) *Ohio State Journal on Dispute Resolution* 349, 379.

39 Maurits Barendrecht, 'Rechtwijzer: Why Online Supported Dispute Resolution is Hard to Implement', *Law Technology and Access to Justice* (Blog Post, 20 June 2017) <https://law-tech-a2j.org/odr/rechtwijzer-why-online-supported-dispute-resolution-is-hard-to-implement/>.

40 Lois R Lupica, Tobias A Franklin and Sage M Friedman, 'The Apps for Justice Project: Employing Design Thinking to Narrow the Access to Justice Gap' (2017) 44(5) *Fordham Urban Law Journal* 1363, 1376.

41 Gideon, *Gideon* <https://www.gideon.legal/>.

42 Robot Lawyer Lisa, *LISA* (2020) <https://robotlawyerlisa.com/>.

43 Promoting Justice Openness, China Court Mobile TV App now go live <https://www.chinacourt.org/article/detail/2015/02/id/1558524.shtml>.

44 Qiang Zhou, Promoting Rule of Law by Telling Stories of Rule of Law in an Innovative Way <http://www.court.gov.cn/zixun-xiangqing-13541.html>.

45 Jena McGill, Suzanne Bouclin and Amy Salyzyn, 'Mobile and Web-based Legal Apps: Opportunities, Risks and Information Gaps' (2017) 15 *Canadian Journal of Law and Technology* 229, 251.

46 Legal Services Commission of South Australia, *LegalAidSA* (2020) <https://play.google.com/store/apps/details?id=com.andromo.dev86688.app95890>.

47 Jena McGill, Suzanne Bouclin and Amy Salyzyn, 'Mobile and Web-based Legal Apps: Opportunities, Risks and Information Gaps' (2017) 15 *Canadian Journal of Law and Technology* 229, 230.

48 Jena McGill, Suzanne Bouclin and Amy Salyzyn, 'Mobile and Web-based Legal Apps: Opportunities, Risks and Information Gaps' (2017) 15 *Canadian Journal of Law and Technology* 229, 237.

49 Jena McGill, Suzanne Bouclin and Amy Salyzyn, 'Mobile and Web-based Legal Apps: Opportunities, Risks and Information Gaps' (2017) 15 *Canadian Journal of Law and Technology* 229, 238.

50 Rangefindr, *Rangefindr* (2019) <http://www.rangefindr.ca/>; Jena McGill, Suzanne Bouclin and Amy Salyzyn, 'Mobile and Web-based Legal Apps: Opportunities, Risks and Information Gaps' (2017) 15 *Canadian Journal of Law and Technology* 229, 238.

51 Blue J Legal, *Blue J Legal* (2020) <https://www.bluejlegal.com/>; Jena McGill, Suzanne Bouclin and Amy Salyzyn, 'Mobile and Web-based Legal Apps: Opportunities, Risks and Information Gaps' (2017) 15 *Canadian Journal of Law and Technology* 229, 238.

52 Jena McGill, Suzanne Bouclin and Amy Salyzyn, 'Mobile and Web-based Legal Apps: Opportunities, Risks and Information Gaps' (2017) 15 *Canadian Journal of Law and Technology* 229, 239–40.

53 Jena McGill, Suzanne Bouclin and Amy Salyzyn, 'Mobile and Web-based Legal Apps: Opportunities, Risks and Information Gaps' (2017) 15 *Canadian Journal of Law and Technology* 229, 239.

54 See Adam Feibel, 'Do-It-Yourself Divorce Law Made Easier: Ottawa Firm Launches New App', *Ottawa Business Journal* (29 February 2016) <https://obj.ca/article/do-it-yourself-divorce-law-made-easier-ottawa-firm-launches-new-app>.

55 'Smart court' is a terminology officially raised by the SPC in 2016 with a view to turning China's court system into a highly intelligent one by rolling out the technology use. This initiative was integrated into China' *National Strategy for the Informatization Development* which is available at <https://chinacopyrightandmedia.wordpress.com/2016/07/27/outline-of-the-national-informatization-development-strategy>.

56 In China, mediation is often part of litigation process and is conducted by judges.

57 <https://www.chinacourt.org/article/detail/2018/08/id/3471944.shtml>.

58 <https://m.chinanews.com/wap/detail/zw/gn/2020/01-14/9059341.shtm>.

59 Adam M Gershowitz, 'Criminal Justice Apps' (2019) 5 *Virginia Law Review* 1, 1–2.

60 Adam M Gershowitz, 'Criminal Justice Apps' (2019) 5 *Virginia Law Review* 1, 13.

61 Adam M Gershowitz, 'Criminal Justice Apps' (2019) 5 *Virginia Law Review* 1, 14.

62 See Jena McGill, Suzanne Bouclin and Amy Salyzyn, 'Mobile and Web-based Legal Apps: Opportunities, Risks and Information Gaps' (2017) 15 *Canadian Journal of Law and Technology* 229, 244.

63 American Psychiatric Association, *Why Rate Mental Health Apps?* (2019) <https://www.psychiatry.org/psychiatrists/practice/mental-health-apps/why-rate-mental-health-apps>.

64 American Psychiatric Association, *App Evaluation Model* (2019) <https://www.psychiatry.org/psychiatrists/practice/mental-health-apps/app-evaluation-model>.

65 Anuj P Kapoo and Madhu Vij, 'How to Boost your App Store Rating? An Empirical Assessment of Ratings for Mobile Banking Apps' (2020) 15(1) *Journal of Theoretical and Applied Electronic Commerce Research* 99, 110–1.

66 New Zealand Ministry of Health, *Guidance on Evaluating or Developing a Health App* (2017) <https://www.health.govt.nz/system/files/documents/pages/guidance-evaluating-developing-health-app-oct17-v2.pdf>.

67 Anuj P Kapoo and Madhu Vij, 'How to Boost your App Store Rating? An Empirical Assessment of Ratings for Mobile Banking Apps' (2020) 15(1) *Journal of Theoretical and Applied Electronic Commerce Research* 99, 110.

68 Ithaca Group, *Evaluation of PENDA: A Financial Empowerment App for Women Escaping Domestic and Family Violence* (Final Report, June 2018) 6–7.

69 Sherley Cruz, 'Coding for Cultural Competency: Expanding Access to Justice with Technology' (2019) 86 *Tennessee Law Review* 347, 374.

70 Sherley Cruz, 'Coding for Cultural Competency: Expanding Access to Justice with Technology' (2019) 86 *Tennessee Law Review* 347, 389, 399–400.

71 Fahimeh Abedi, John Zeleznikow and Chris Brien, 'Developing Regulatory Standards for the Concept of Security in Online Dispute Resolution Systems' (2019) 35 *Computer Law & Security Review* 1.

72 Fahimeh Abedi, John Zeleznikow and Chris Brien, 'Developing Regulatory Standards for the Concept of Security in Online Dispute Resolution Systems' (2019) 35 *Computer Law & Security Review* 1.

73 Daniel Rainey, 'Creating Standards for ODR' (2017) (4)2 *International Journal on Online Dispute Resolution* 21, 24.

74 See for example Catherine Han et al. (2020) *The Price Is (Not) Right: Comparing Privacy in Free and Paid Apps.* In: Privacy Enhancing Technologies Symposium (PETS 2020), July 14–18, Montreal available at <http://eprints.networks.imdea.org/2121/>.

75 National Alternative Dispute Resolution Council (NADRAC), *A Framework for ADR Standards* (Attorney-General's Department, Canberra, 2001) 13–4.

4 Justice apps in context

Introduction

This chapter explores the use of justice apps in two main contexts – the family law and criminal law areas – in addition to the increasingly popular use of apps that support video conferencing in light of the COVID-19 pandemic. The end user perspective is also considered. In this regard, the authors note that often courts collect little demographic information about court users, although in the criminal area there is generally more information available.[1]

The family law and criminal law areas have been selected for specific analysis as each area raises unique app development issues and has also been the focus of previous research and commentary. In this regard, the very different issues that surface in relation to each area highlight some of the matters that are relevant to the ongoing evaluation of justice apps more generally and reflect the need to consider additional specific app objectives in the context of jurisdictional objectives. For example, an objective of a family law app may include supporting the best interests of a child which may in turn require that additional assessment or evaluation criteria are developed to ensure that this objective is met by the app.

Family law

In Australia, Sourdin and Liyanage have noted that the use of ODR for the resolution of family law disputes has been supported and is likely to develop further due to two factors: (i) the development of laws which require most family disputes to be mediated through a family dispute resolution process as a mandatory requirement before resorting to the courts; and (ii) increases in the use of technology in both formal courts and informal ADR dispute resolution processes during family disputes.[2] This can be contrasted with earlier research which found that ODR for family disputes had not received as much critical attention as the application of ODR to other areas,[3] with a dated 2004 survey of 115 ODR sites finding that only five dealt specifically with family disputes.[4]

Greater use of ODR in the family law context can have a number of important benefits. As outlined by Bell, these include time and cost savings, control and ownership of the outcome, and the preservation of relationships.[5] In the Netherlands, an evaluation of *Rechtwijzer* found that 82 percent of surveyed users felt 'respected' or 'very respected' by lawyers and/or mediators on the platform which sought to maximise lawyers' interventions in such a way as to aid users but not supersede their judgement.[6] In terms of cost savings, Tyler and McPherson have noted that processes surrounding separation and divorce, especially in relation to parties that are geographically remote, can involve expensive correspondence and litigation, and greater than normal costs in time, travel, and accommodation.[7] Given that the financial settlement in a divorce in countries such as Australia comes out of the one pool of assets, any process that reduces costs is likely to be of benefit.[8]

Improved access to justice in the family law context is also a key consideration. Access to justice in family law matters has been identified as a serious problem in many countries such as Australia. As noted by Bell, family law has historically been an area that many people end up traversing with only limited legal assistance.[9] Not only does government funding of legal aid continue to decline, but a large proportion of people who are unable to afford the cost of engaging a lawyer do not qualify for legal aid.[10] Higgins terms this the 'missing middle' of the legal services market.[11] As noted by Bell, it is this 'missing middle' who are the expected or intended beneficiaries of the use of automated systems in the family law context.[12]

'Self-help' or 'non-lawyer' options are not new to family law. Bell has noted that information about family law has been on the Internet for many years, and has already produced a cultural change toward self-help.[13] Crowe et al have similarly observed that people are now more likely to seek information on the Internet, including in areas which would have once been considered to require professional advice.[14] According to Bell, arguments against the sophisticated ODR options now available are essentially the same as those raised about the proliferation of online self-help information.[15] Nevertheless, despite the volume of information available, non-lawyers seeking family law information in the online environment reportedly find it difficult to traverse its complexities, and hard to evaluate the credibility of different sources.[16] As noted by Bell, 'the potential benefit, then, of using automated tools is to more precisely direct non-lawyers to relevant information'.[17]

At a basic level, technology can assist people by putting them in touch with a lawyer or facilitate the completion of relevant paperwork to finalise a dispute. This type of 'supportive technology'[18] does little to differentiate between individual disputes. For example, a website based in the USA – *OurDivorceAgreement* – enables couples seeking an uncontested divorce to complete their divorce agreement online, providing forms for filing with a court.[19] In Australia, the online portal *Settify* allows potential clients to provide instructions online prior to their first face-to-face meeting with a lawyer.[20] The *LegalAidSA*[21] and *PENDA*[22] apps discussed in Chapter 3 also operate by providing legal information and resources, with the latter specifically focussed on the family law sector and providing access to legal, financial, and safety information for victims of domestic and family violence.

Justice apps can also assist separated parties with scheduling and communication. In their 2019 study of smartphone apps available in Australia for separated or divorced families, Smyth and Fehlberg found that the largest group of apps ($n = 25/43$) comprised of integrated collaboration software systems based around an electronic calendar for managing children's schedules.[23] One such app is *Our Family Wizard*, developed in the U.S.. *Our Family Wizard* allows parents to store their schedule, files, contacts, and communication all within one app, allowing them to 'solve shared custody challenges faster and without confusion'.[24] Lawyers, mediators, and other family law professionals can also use the platform to review client communications, download reports, and better manage client relationships.[25]

In Australia, the *MyMob* app performs a similar function.[26] Smyth and Fehlberg also identified four apps that were specifically focussed on improving communication between parties to a family law dispute.[27] An example is *Amicable Divorce* in the UK which is stated to help parties 'build a dialogue' to support co-parenting, and not leave them 'bitter or dependent on lawyers to sort things out'.[28] Other apps identified by Smyth and Fehlberg were directed at record-keeping and keeping track of financial expenses and payments (e.g., *My Divorce Notes*[29]), or providing information and linked access to support resources.[30] Reporting on the financial cost of such apps, Smyth and Fehlberg noted that this ranges widely, from free (e.g., *MyMob*) to $99 AUD per annum (*Our Family Wizard*).[31]

Family law apps and dispute resolution

More sophisticated apps are capable of providing a dispute resolution process that is tailored or personalised to the individual dispute. At this level, the technology may either replace activities and functions that were previously carried out by humans ('replacement technology'), or provide for very different forms of ADR, particularly where processes change significantly ('disruptive technology').[32] An example is the development of negotiation support systems for the resolution of family law disputes. Programmes such as *Split-Up*, *FamilyWinner*, and *AssetDivider* use simplified AI processes. *Split-Up* employs branching technology to assist parties in calculating the division of property in family law proceedings. It is an artificial intelligence (AI) system which searches a database of Australian family law decisions to give an indication of the likely result of a particular negotiation demand. The system has determined that there are 94 factors relevant for a percentage split decision.[33] *FamilyWinner* and *AssetDivider* employ decision analysis techniques and compensation/trade-off strategies to facilitate the resolution of family disputes.[34] Concerns have been raised, however, that such systems may ignore issues of justice, given that the outcome should reflect the needs of the children rather than the interests of the parents.[35]

Internationally, from 2014 to 2017, *Rechtwijzer* provided separating couples in the Netherlands with an online process to resolve their disputes. The platform would identify points of agreement between the parties and propose solutions. If parties found they could not effectively communicate or reach a decision on their own, they could request either an online mediator to support negotiations or a neutral party to issue a binding decision.[36] Funding for the *Rechtwijzer* programme was discontinued after it was deemed financially unsustainable after running for three years.[37] It has been reported that approximately 60 percent of those who used the platform proceeded through to finalising an agreement and registering it.[38]

Notably, research has shown that until recently there were very few apps targeted at divorced or separated individuals which had been developed in Australia, with the majority developed in the U.S. and directed at the U.S. legal context.[39] It has also been reported that there is a limited 'buy-in' culture amongst parents and legal professionals when it comes to justice apps designed to assist separated parents in managing their parenting. This suggests a gap between developer and/or government promotion of the increased use of technology in the family law context, and the end-users of such technologies.[40]

The *Adieu* app in Australia has been developed to meet this need. The authors, who have been involved in reviewing this app, considered research relating to the use of the app by users and professionals.[41] As Sourdin explains, *Adieu* Technologies aims to provide cooperative and legally sound post-divorce financial settlements and parenting agreements through its app.[42] The app combines self-service with human assistance and expert determinations. '*Lumi*' – which forms part of *Adieu* – is a 'personal separation guide' that is supported by AI.[43] '*Lumi*' provides real-time feedback and guidance, while *Adieu's* professional customers (viz. accountants, financial advisers, lawyers, and mediators) support clients to reach an agreement via a web-based interface. Once an agreement is reached, the system generates an application that can be filed with a court to make the agreement legally binding.[44]

Interestingly, many users, even those with limited exposure to apps, were positive about app use. For example, one participant in the *Adieu* research project responded that 'I'm pretty new to apps but am learning. They're not so bad, but don't really replace people. On the plus side, they're neutral and don't judge you!' Similarly, one participant commented that 'some [apps and online technologies] are great, some are just a waste of time'.

For many users, the *Adieu* processes were supportive and helpful. Although the user survey sample was relatively small ($n = 37$), the results indicate that supportive and interactive apps can be of assistance to separating couples.

An additional app called *Amica*, was recently launched by the Australian Government (instead of the private sector) which helps separating couples to make parenting arrangements and divide their assets 'amicably'.[45] *Amica* can make suggestions about dividing money and property based on the information that the separating couple enters in conjunction with the consideration of relevant legal principles,[46] however, does not include triage services or more specific client support.

ODR processes in the family law context can allow the parties to avoid face-to-face interaction and 'divorce at a distance'.[47] This is particularly useful where there are allegations of violence or abuse, or significant power imbalances which mean that traditional face-to-face forms of ADR are inappropriate.[48] Indeed, Sourdin and Liyanage have suggested that ODR is especially suitable for family law disputes.[49] Bell has also recognised that 'the complete physical (and possibly temporal) separation of the parties in particular lends itself to family mediation or family dispute resolution (FDR), especially in cases involving allegations of violence'.[50]

Further, Bellucci, Venkatraman, and Stranieri highlight technology's ability to filter negative emotions out of the dispute process, noting that this provides another reason ODR has been embraced in the family law context.[51] On the other hand, Condlin has argued that the removal of face-to-face interaction can 'suspend, at least in part, the felt obligation to be sociable' meaning ODR, in an 'ironic twist', 'might undo some of the important reforms produced by the ADR movement of the past several decades'.[52]

When it comes to the potential role of justice apps in this context, empirical evidence has suggested that the higher the conflict between separating parents, the more likely they are to communicate via text and email rather than utilising wider smartphone features. In their 2019 study of the smartphone apps available for separated or divorced families in Australia, Smyth and Fehlberg found that in high-conflict separations, apps have the potential to increase conflict, with app usage more likely to occur where parents are already cooperating. In other words, 'those most in need of the mechanisms for reducing and improving communication that digital technologies can offer also [appear] to be the least likely to make full use of those technologies'.[53] Scheduling apps, for instance, were viewed by parents as a more effective tool once a care pattern was already established.[54]

Video conferencing

Related to questions concerning the use of family law apps are apps that enable those in a dispute to deal with one another remotely. In this regard, there is some domestic and international research exploring the effectiveness of video conferencing in ADR and legal proceedings more broadly. 'Video conferencing' can be defined as 'all synchronous (two way) communication with audiovisual interface, whether via integrated service digital network ("ISDN"), satellite or internet protocol ("IP") with video conferencing technologies'.[55] In recent years, growth in the use of video conferencing has been attributed to a reduction in the cost of technology, increased accessibility, and the diminishing feasibility and high cost of maintaining resident legal services in some rural, regional, and remote areas.[56] In 2020, the COVID-19 global pandemic[57] resulted in a massive increase in the use of video conferencing, with some accompanying concerns[58] relating to effectiveness, security, and potential negative impacts on justice.

The increase in what Forrell, Laufer and Digiusto[59] describe as video conferencing with an all synchronous (two way) communication with audio-visual interface includes the use of web-based platforms such as *Teams, Skype, Zoom, Google Hangouts,* and *WebEx.* Whilst some video conferencing can be integrated with justice app functions, often this is not the case. As Zeleznikow has noted, a truly helpful ODR system should provide the following six facilities[60]: case management, triaging, advisory tools for reality testing,[61] communication tools,[62] decision support tools,[63] and drafting software.[64] Further:

> With citizens of many (if not all) communities forced into isolation due to COVID-19 restrictions, litigants are no longer meeting face-to-face. The justice system needs to operate in these circumstances – especially so in cases of family disputes and bail applications. However, the authors note that the systems currently in use, such as Immediation,[65] MODRON,[66] and Our Family Wizard[67] only offer two out of the six essential facilities of Zeleznikow's ODR model, viz. case management and communication.[68]

There is some existing research that addresses the effectiveness, benefits, and challenges associated with the use of video conferencing, and many of these findings are useful when it comes to assessing the appropriateness of ODR processes in the family law context.

First, video conferencing can facilitate lawyer-client communication. In NSW, Australia, video conferencing has been extensively used for the provision of legal advice to prison inmates. In 2009–2010, over 10,000 video conferences were held between lawyers and their clients in NSW correctional centres, compared to only 938 video conferences between 2003 and 2004.[69] Video conferencing is often assumed to provide a preferable mode of communication when compared with other technologies that do not allow participants to see each other 'face-to-face', and thus cannot facilitate nonverbal communication. For instance, lawyers have reported that video conferencing is beneficial in allowing them to observe client responses.[70] Further, many video conferencing systems allow for the efficient viewing and exchange of documents.[71] Of course, in the family law context, video technology which allows 'face-to-face' communication between separated parties may be undesirable in cases where there are significant power imbalances or allegations of violence and/or abuse.[72]

Second, cost savings are often identified as a further benefit associated with the use of video conferencing, particularly in terms of the reduced time and costs associated with travelling to remote locations.[73] In 2018, a one-month pilot programme involving online hearings in 65 cases in the Victorian Civil and Administrative Tribunal (VCAT) in Australia found that the use of video conferencing facilities resulted in a higher respondent participation rate than is usual in small claim dispute resolution, greater convenience for the parties, easier submission of evidence and time savings.[74] Nevertheless, there is a dearth of research exploring the costs associated with video conferencing – including set up, maintenance, and support costs – making it difficult to determine how video conferencing can provide a more cost effective alternative to face-to-face legal assistance.[75] The COVID-19 pandemic has highlighted that such systems can be implemented at relatively low cost, although bandwidth capacity and user technology issues remain significant obstacles (see discussion relating to the digital divide in Chapter 5).

There has been some relevant research in relation to the challenges associated with the use of video conferencing technology in legal proceedings. First and foremost, technological difficulties can reduce the effectiveness of communication by video conference. It has been reported that in video conferences between client and solicitor, technological difficulties can be frustrating for both parties, and result in the need for extra appointments to complete the legal assistance.[76] Convenience has been identified by the Law and Justice Foundation of New South Wales as another challenge associated with the use of video conferencing.[77] In rural, regional, and remote areas, a lack of access to video conferencing technology can be a particular barrier for clients.[78]

Privacy and confidentiality concerns and user comfort with the technology are two further barriers. In relation to the former, it has been reported that clients who have to access video conferencing technology in potentially non-confidential locations can be concerned about 'being seen to have a problem' by members of their community.[79] Ebner highlights an additional privacy-related concern: the fact that in ODR processes, parties can never know who else is 'in the room', potentially leading to reduced information-sharing.[80] In the context of the COVID-19 pandemic, this issue has been raised in relation to family law conferencing and has been a particular concern where children may have been exposed to parental video conferencing whilst in home isolation.

The risk of a party surreptitiously recording a video conference is a further concern.[81] Questions relating to both privacy and the confidentiality of video conferencing have resulted in some changes in 2020 to a range of video conferencing products and services. For example, 'Zoom bombing' that arose as a more specific issue during the pandemic led to some significant changes to the Zoom platform.[82] Notably the risk may not only be related to the app, as it may be the mobile phone or computer that is hacked (rather than the app itself).

In relation to a client's comfort using video conferencing technology, the Australian Pro Bono Centre has reported that the experience of those involved in video conferencing pilot projects suggests that some people will need to have a support person who can both assist with the use of the technology, and help the client to understand the advice and complete any follow-up actions.[83] Whilst at first glance user comfort and/or a lack of support may therefore be a barrier to the use of video conferencing, the Law and Justice Foundation of New South Wales has reported that although clients may initially be reticent, 'they generally adapt and become comfortable enough with the technology'.[84] Despite these findings, the reality in the COVID-19 pandemic is that video conferencing has been introduced, often with little user engagement or preparation.

There are particular issues where people have disabilities with a recent UK report noting that '[v]ideo hearings can significantly impede communication and understanding for disabled people with certain impairments, such as a learning disability, autism spectrum disorders and mental health conditions'. Such issues may raise more significant issues in some justice areas, with the report noting that '[p]eople with these conditions are significantly over-represented in the criminal justice system'.[85] Other more recent research has suggested that good design and the implementation of guidelines, that include having supportive trial runs, can assist people to use video conferencing more effectively (See discussion on 'the digital divide and accessibility' in Chapter 5).[86]

There are also concerns about the effectiveness of video conferencing.[87] As noted by the National Alternative Dispute Resolution Advisory Council (NADRAC) in Australia, although video conferencing provides an approximation to face-to-face interaction, 'images are two-dimensional, and, as eye contact is via a fixed camera, some information gained from eye contact is lost'. In addition, 'lagging can create delays in responses and lead to a perception of hesitancy'.[88] Tan has also highlighted the lack of a sense of warmth and empathy or a 'personal touch' between the parties.[89]

Significantly, the information lost as a result of using such technology 'may have an impact on some of the intangible aspects of human relationships', such as making it difficult to create trust.[90] Ebner and Thompson explain that one way trust is created in face-to-face interactions is through postural mirroring and 'unconscious mimicry' – the repeating of another's nonverbal behaviour. Both of these can be lost in video conferencing, along with the opportunity to create or improve rapport, empathy, and immediacy.[91] This is a significant disadvantage, with empirical evidence showing that in the mediation context, the ability of a mediator to gain a party's trust can be critically important.[92] As video conferencing relies on internet connectivity, the issues can be compounded where one party has no video (for example with only a name or photo appearing) whilst others may have good quality video connections.

In the negotiation context, Ebner has observed that physical distance can create a sense of interpersonal 'otherness' which challenges identity-based trust.[93] A study of the use of video conferencing in ODR similarly found that 'building trust and rapport is essential to resolving disputes'.[94] Similar findings have been reported in the context of the lawyer-client relationship.[95] Interestingly, Condlin has suggested a further way in which face-to-face communication can build trust where video conferencing cannot:

> Wholly apart from its comparative advantage in communicating non-verbal information, face-to-face interaction shows respect for a person in a way that videoconferencing cannot by saying, implicitly, "You're worth the expense and a few days of my time."[96]

Whether this is a realistic criticism is questionable, however, the issues that relate to the place where video conferencing occurs are relevant. For example, video conferencing when conducted from a home environment can reveal much about the party connecting via video conferencing. Whilst some may access mock up photos to alter backgrounds, for many, video conferencing could be an embarrassing and unsettling experience.

Considering the above challenges, it is perhaps unsurprising that the Law and Justice Foundation of New South Wales in Australia highlighted the need for caution in the large scale roll out of video conferencing technology for legal assistance.[97] It has also been reported that lawyers and clients prefer in-person meetings to video conferencing, with the latter only seen as 'an acceptable and functional interface for clients and lawyers when in-person meetings were not possible'.[98] In the ADR context, the NADRAC has suggested a number of strategies for making video conferencing more effective, including explaining to parties the limitations of the medium.[99] Despite this, however, and in view of COVID-19 isolation requirements, video conferencing has become the norm in many courts and tribunals around the world.[100]

It is also important to note that some of the challenges identified above are in fact better construed as benefits in the family law context. Here, the loss of 'information' such as non-verbal cues may in fact be useful, particularly where 'interpersonal dynamics are destructive' and/or there is 'a history of physical intimidation or enmeshed conflict'.[101] Finally, not all research has drawn a link between video conferencing and decreased levels of trust between participants. In their study of mediators engaged in ODR, Exon and Lee reported no statistically significant difference in the levels of trust between participants who interacted with a mediator face-to-face, and those who participated in a video-collaborated mediation.[102]

A number of useful insights can also be drawn from research on the use of video conferencing during court proceedings. Here, trust issues again pose a significant challenge. In research exploring the use of video conferencing in immigration proceedings, Haas has reported that defendants suffer from a loss of trust and credibility.[103] This is seen as a particular disadvantage in immigration hearings which often involve few witnesses, and an outcome which can turn on the perceived credibility of the defendant.[104] Hart has identified a further challenge in this context: a 'loss of solemnity' of the courtroom context due to the lack of physical presence.[105] Similarly, Tan has highlighted how the erosion of immediacy which flows from the use of video conferencing facilities can also erode perceptions of the importance of the dispute and its resolution.[106]

In terms of family law hearings, Foster and Cihlar have identified an additional risk: the fact 'there is less opportunity to observe if witnesses are testifying under any form of impairment, if they are being coached off-screen, or if they are reading from notes that are not part of the record'.[107] In the context of civil trials, it has again been noted that notwithstanding technical improvements in video technology, there remains a risk of significant unintended effects on both credibility assessments and the emotional connections.[108]

Criminal law

Numerous apps have been developed in the criminal justice system. Whilst some have been directed at offenders, many have been directed at law enforcement agencies and courts. As with family law, the COVID-19 era has heralded a new wave of video conferencing apps that have been used differently by court systems around the world and which have raised concerns about the types of matters that should be dealt with via video conferencing.[109] Currently, there are few apps that are directed at lawyers in the criminal justice area.

A number of commentators suggest that justice apps can support the justice system and result in a fairer system. For example, Gershowitz has noted that criminal justice apps can 'democratize' the criminal justice system by teaching individuals about the law, helping suspects and defendants connect with lawyers, helping defendants navigate confusing court systems and even seeking to bring systemic changes to the criminal justice system.[110]

In the criminal law context, apps can also provide information about legal rights and help defendants navigate the court system. In this sense, apps can thus assist the public as a 'supportive' technology. One example is the aforementioned, free-to-download, *NSW Pocket Lawyer App*, developed by Sydney Criminal Lawyers, which offers information about common criminal and traffic offences in a user-friendly, plain-English setting.[111] Similar legal advice apps for criminal cases include the *Go To Court App* in Australia, in relation to which it has been commented:

> This Australian first-of-its-kind App will help people get the answers they need in real time and provide relief in stressful situations. It really is an App that puts a lawyer in everyone's pocket.[112]

In terms of video conferencing apps, a range of specific concerns have been raised in criminal law proceedings. There is again evidence in this context that interaction through the barrier of technology can make a defendant harder for the audience and adjudicator to relate to.[113] Treadway Johnson and Wiggins have also noted that a defendant's nervousness because of the presence of a camera may negatively affect a judge's perceptions of the defendant's credibility.[114] Credibility can also be affected by the transmission of auditory information through videoconferencing, with high voice frequencies which convey meaning about the emotional state of the speaker partly excluded when such technology is used. This is problematic, because it is precisely this information that may be critical to judgments about the defendant's remorse and/or credibility.[115]

Various empirical studies support the above concerns. A study of bail hearings conducted via closed circuit television in Illinois between 1999 and 2009 found that defendants were significantly disadvantaged by video-conferenced bail proceedings, due to a 'dehumanization' that encouraged a harsher response than occurred when the judge was faced with a live individual.[116] A study of decisions in asylum hearings between 2004 and 2005 similarly found that individuals who had in-person hearings were almost twice as likely to be granted asylum as those whose hearing was held by video-conference.[117]

More recent research has suggested that, as with family law video conferencing, some concerns can be addressed by having a range of systematised approaches and adequate video conferencing.[118] However, the authors of that recent study noted that their findings were limited to non-criminal matters and that the concerns in the criminal area may be significantly different.

Technologies that are used for purposes of pre-trial risk assessment, sentencing, or assisting judges with decision-making more generally are more disruptive by nature and can raise important due process concerns and associated bias issues. One such example is an AI tool based on algorithms in the U.S. called the *Correctional Offender Management Profiling for Alternative Solutions (COMPAS)*, which has been used in sentencing offenders. The issues arising out of this system were examined in *State v. Loomis*.[119] In that case, the defendant had pleaded guilty to two charges relating to a drive-by shooting and *COMPAS* recommended the maximum sentence, which the offender received. As *COMPAS* was proprietarily protected, the examination of its software was prevented by the Wisconsin Supreme Court and later the U.S. Supreme Court. This issue of who owns an app, who owns data and how a court or user can explore outcomes, has been the subject of some recent research in the justice area and raises specific concerns in the justice app context.[120]

In the COMPAS/Loomis matter, some testing of the tool suggested that black defendants were twice as likely to be labelled high risk when compared with white defendants and that white defendants who had been categorised as lower risk were more likely to reoffend.[121] This concern has been raised in the context of algorithmic bias, where it has been suggested that apps which are not well developed can be 'opaque' and trigger biased outcomes.[122] Another study by Dressel and Farid, found that the tool *COMPAS* used to predict recidivism was as accurate as an online poll of 400 random people with no criminal or legal training (67 percent accuracy compared with 65 percent accuracy for *COMPAS*).[123]

A later study by Lin found that the *COMPAS* algorithm could perform better than a human under some circumstances and that it was relevant whether or not feedback was given to those surveyed about whether the person had reoffended. In replicating the study above, Lin removed this feedback mechanism as 'in real life, it can take months or even years before criminal justice professionals discover which people have reoffended'.[124] Where the feedback was retained, results were similar to those in the original Dressel and Farid study.[125] However, after removal of the feedback, the researchers found that the gap in accuracy significantly favoured the algorithm.[126]

A more recent case in Kansas, U.S., concerns information offenders are given—or not—about their risk scores as a result of the court's application of criminal justice algorithms.[127] In one matter, an applicant pled no contest to a criminal threat charge. He was then evaluated using the LSI-R (Level of Service Inventory-Revised) risk assessment tool. When he asked to see the results, he was prevented from seeing the specific questions and answers and the scores associated with those questions. Rather, access was only granted to a cover page that summarised general scores. After he was sentenced by a district court to a highly supervised form of probation generally used for moderate or higher-risk offenders, he challenged his sentence before the Kansas Court of Appeal, arguing that the refusal to disclose the details of his LSI-R assessment violated his right to due process. The appeals court ruled in favour of the applicant, noting that denying the applicant access to his complete LSI-R assessment made it impossible for him to challenge the accuracy of the information used in determining the conditions of his probation.[128]

Perhaps realising the potential for algorithms to undermine a 'just' outcome to offenders, some prominent groups in the U.S., such as the Pretrial Justice Institute (PJI), a previous strong advocate for the introduction of these tools, have recently changed their position on pretrial risk assessment tools.[129]

In February 2020, PJI released its '*Updated Position on Pretrial Risk Assessment Tools*', where it stated:

> We now see that pretrial risk assessment tools, designed to predict an individual's appearance in court without a new arrest, can no longer be a part of our solution for building equitable pretrial justice systems. Regardless of their science, brand, or age, these tools are derived from data reflecting structural racism and institutional inequity that impact our court and law enforcement policies and practices. Use of that data then deepens the inequity.[130]

In comparison to the more extensive critical reflections in the U.S. on the use of AI and other technologies in criminal cases, Chinese courts are keen to continue the construction of a smart court system where various technologies (including AI and big data) are utilised. Such technologies include apps designed to support 'uniformity in law' and are directed (to some extent) at reducing human bias. For example, Justice Yadong Cui of the Shanghai High People's Court has noted that:

> Because the judicial personnel are different individuals with subjective initiative, there will inevitably be some differences in ensuring uniformity of law, which will result in inconsistent law

enforcement and different judgments in cases sharing the highly similar facts. Application of artificial intelligence can provide relatively streamlined judicial reasoning and evaluation standard, provide the judge with all similar cases, laws, regulations and judicial interpretations and so on,[131] so the judge can strictly follow the rule of evidence and procedure, which will reduce judicial arbitrariness and promote justice.[132]

Such apps extend well beyond apps used in most other countries and may be interconnected with other data from outside courts and other systems. It is also clear that they are directed more clearly at courts and judges. For criminal cases, Cui strongly promotes an online platform called '*Shanghai Intelligent Auxiliary System of Criminal Case Handling*' where mass judicial data (e.g., case reports, repository of legislation, and judicial interpretations) is collected from all courts in the country and used by judges in Shanghai to assist with their decision-making process to ensure their judgments for similar cases are in line with those delivered by their counterparts in the rest of the country.[133]

Despite the seemingly promising application of newer technologies in China's courts, some scholars have raised concerns related to the use of such technologies. For example, Zuo contends that in China, the quantity and quality of judicial data that have been relied on by developers to provide repositories for judges to facilitate their decision-making is questionable.[134] Specifically, Ma, Yu, and He comment that only 50 percent of the total court judgments in China had been moved online and this has been used to inform the data approach. With nearly half of all judgments unavailable to app developers, it is questionable how reliable any suggested outcomes may be.[135] In addition, as Sourdin has noted, there are significant issues in relying only on data from cases that have been heard by judges, as such cases may be somewhat aberrant in that 'easier' cases may have been settled[136] and there are also judicial independence and ethical issues that could be raised in some jurisdictions by more extended approaches.

Conclusions

As noted above, there is some research in both the family law and criminal justice areas that suggests app development can be problematic and that some apps may further disadvantage those who are already vulnerable. This matter is discussed more fully in Chapter 5 in the context of the digital divide. In addition, the opacity of some apps can mean that a review of an app recommendation or finding is unlikely to occur. Clearly such issues are more problematic with apps that are not merely supportive. However, even with supportive apps in the video conferencing area, there may be issues that compound existing inequities. As noted in Chapter 3, there is also little evidence that some justice apps have been designed within a framework where such matters are thoughtfully considered. In addition, few justice apps that are in operation have been evaluated.

Despite this, there is now some robust evidence that shows that many people can be satisfied with systems and apps that support justice services. In particular, past research has revealed that many users in the non-criminal area have generally positive attitudes towards and experiences with justice apps.[137] In comparing past research to the results of the *Adieu* study, it seems clear that there are strong thematic consistencies between users' experiences with justice apps. For example, as outlined above, an evaluation of *Rechtwijzer*, an ODR platform for separating couples in the Netherlands, found that 82 percent of users felt 'respected' or 'very respected' by lawyers or mediators on the platform.[138] These same user attitudes and feelings of respect towards family law professionals are reflected in the results of the *Adieu/Lumi survey*, with 68 percent of participants indicating that they were comfortable with *Lumi* introducing them to human professionals.

As noted above, the *Rechtwijzer* evaluation also revealed positive results for user empowerment, with 84 percent of surveyed participants perceiving they had increased control over their separation from using the platform.[139] These perceptions of empowerment were also reported by *Adieu/Lumi* users. That is, 87 percent of participants agreed that *Lumi* helped them better understand the separation process and what would be required of them to continue with their separation. Indeed, perceptions of empowerment and increased control over the separation process were revealed to remain with participants even after the divorce process was finalised.[140]

Further, over half the participants in the *Rechtwijzer* evaluation reported experiencing 'low' or 'very low' stress levels during their separation.[141] These low feelings of stress were also reported by *Adieu/Lumi* users. However, each of these apps relies on and enables referral to human beings who can provide support at various stages of a dispute. In general, it could be suggested that apps without such a feature (including video conferencing apps) may not be likely to satisfy a user. In addition, an app that makes a decision or determination, or contributes significantly to a determination, may be viewed with suspicion and concern particularly if human oversight appears to be limited.

Where apps are integrated within a court or tribunal functions, as with the CRT in Canada that is discussed above, high levels of satisfaction have been recorded, with 79 percent of people using the CRT reporting they were likely to recommend the CRT to others.[142]

At the same time, however, as ODR may never be able to completely replace face-to-face contact with a lawyer, it is likely that many apps will continue to provide a referral structure and support more effective triaging so that time spent with lawyers and other experts (such as financial advisers and mediators) is supported by 'wise' referral. In this regard, the use of targeted justice apps rather than general justice apps can enable more people with legal needs to have access to advice and to see a lawyer when required.[143] Human lawyers can then formulate their conclusions or provide advice on how to respond[144] (see also discussion in Chapter 6 relating to the unbundling of legal services).

Justice apps that are more focussed on the giving of legal advice with no or little human interface are more problematic (as are similar apps in the health field that have in the past raised misdiagnosis concerns[145]). A number of writers have indicated that automated options should not be viewed as appropriate substitutes for professional family lawyers, particularly in the case of vulnerable clients and children.[146] This is not only because such automated responses may be inaccurate but because, according to Bell, family law involves emotional work on the part of the lawyer because clients usually seek, and require, more than 'pure' or mechanistic legal advice.[147]

That is, family law is often seen as necessitating skills which are not strictly technical or legal, but rather, which fall into the category of 'life skills' which are attained through experience rather than formal training.[148] Scheiwe Kulp has also noted that parties are often interested not just in the dispute at a 'technical' level, but also at a 'personal level'.[149] Scheiwe Kulp argues that 'if a party having her "day in court" is a substantial factor in achieving satisfaction, then a process that involves little or no face-to-face contact, as in most ODR, might be considered inadequate.'[150] Finally, Smyth and Fehlberg in their study of apps available in Australia for separated or divorced families also report that parents often view smartphone features as an inappropriate or inadequate substitute for human interaction.[151] One parent stated the following:

> I'm not sure an app ... can replace the need for interpersonal skills and/or support and mediation required to come to parenting arrangements. I think it is critical that both parents feel heard and listened to and validated. I'm not sure an app can do that.[152]

International research also supports this conclusion. In her Canadian study of self-represented parties, Macfarlane noted that many 'expressed the need for more than on-line resources, however good', instead requiring 'human contact and support as they navigate the justice system and prepare their case to the best of their ability'.[153] North American scholars have also suggested that lawyers practising in family law will continue to enjoy greater job security when compared to their colleagues in other areas of the law, given the importance of human interaction for family law clients.[154] In the Netherlands, Dijksterhuis identified the following possible reason for the discontinuation of *Rechtwijzer*:

> Rechtwijzer uit Elkaar might have been too innovative and therefore too uncertain for people in a divorce situation; it was perhaps a bridge too far to expect that huge numbers of people would go for it in such a short period, when in a stressful and unknown situation ... [ODR] is still an unknown; people do not trust it yet, especially in a complex family matter as a divorce. They seem to prefer the safe and familiar route through lawyers.[155]

Beyond the need for human interaction, there are other reasons why justice apps may not be able to entirely replace face-to-face contact with a lawyer. Giddings and Robertson, for instance, highlight 'the highly emotional nature of family disputes and the aggression, depression and anger that is associated with some such disputes'.[156] Against this backdrop, it is suggested that 'it doesn't matter how hard you work with the clients skilling them up, they are not capable of doing a whole lot of things'.[157] One parent in Smyth and Fehlberg's study also reported that apps 'remove the humanity from the intended purpose' when it comes to working out parenting arrangements, noting that 'most loving parents would think this way'.[158]

Notes

1 See Tania Sourdin, Naomi Burstyner, Chinthaka Liyange, Bahadorreza and John Zeleznikow, 'Using Technology to Discover More About the Justice System', (2018) *Rutgers Computer and Technology Law Journal*, 44(1) 1–32.
2 Tania Sourdin and Chinthaka Liyanage, 'The Promise and Reality of Online Dispute Resolution in Australia' in Mohamed S Abdel Wahab, Ethan Katsh and Daniel Rainey (eds), *Online Dispute Resolution: Theory and Practice a Treatise on Technology and Dispute Resolution* (Eleven International Publishing, 2012) 483, 484–5.

3 Melissa Conley Tyler and Mark McPherson, 'Online Dispute Resolution and Family Disputes' (2006) 12(2) *Journal of Family Studies* 165, 170.

4 Melissa Conley Tyler, '115 and Counting: The State of ODR 2004' (Proceedings of the Third Annual Forum on Online Dispute Resolution, University of Melbourne, 2004).

5 Felicity Bell, 'Family Law, Access to Justice, and Automation' (2019) 19 *Macquarie Law Journal* 103, 129.

6 Maurits Barendrecht, 'Rechtwijzer: Why Online Supported Dispute Resolution Is Hard to Implement', *Law Technology and Access to Justice* (Blog Post, 20 June 2017) <https://law-tech-a2j.org/odr/rechtwijzer-why-online-supported-dispute-resolution-is-hard-to-implement/>.

7 Melissa Conley Tyler and Mark McPherson, 'Online Dispute Resolution and Family Disputes' (2006) 12(2) *Journal of Family Studies* 165, 170.

8 Melissa Conley Tyler and Mark McPherson, 'Online Dispute Resolution and Family Disputes' (2006) 12(2) *Journal of Family Studies* 165, 170.

9 Felicity Bell, 'Family Law, Access to Justice, and Automation' (2019) 19 *Macquarie Law Journal* 103, 103.

10 Felicity Bell, 'Family Law, Access to Justice, and Automation' (2019) 19 *Macquarie Law Journal* 103, 113.

11 Andrew Higgins, 'The Costs of Civil Justice and Who Pays' (2017) 37(3) *Oxford Journal of Legal Studies* 687, 692. See also Margaret Castles, 'Expanding Justice Access in Australia: The Provision of Limited Scope Legal Services by the Private Profession' (2016) 41(2) *Alternative Law Journal* 115, 117.

12 Felicity Bell, 'Family Law, Access to Justice, and Automation' (2019) 19 *Macquarie Law Journal* 103, 113.

13 Felicity Bell, 'Family Law, Access to Justice, and Automation' (2019) 19 *Macquarie Law Journal* 103, 109. See also Rosemary Hunter, Jeff Giddings and April Chrzanowski, 'Legal Aid and Self Representation in the Family Court of Australia' (Research Paper, Griffith University Socio Legal Research Centre, May 2003).

14 Jonathan Crowe et al., 'Understanding the Legal Information Experience of Non-Lawyers: Lessons from the Family Law Context' (2018) 27(4) *Journal of Judicial Administration* 137, 137.

15 Felicity Bell, 'Family Law, Access to Justice, and Automation' (2019) 19 *Macquarie Law Journal* 103, 109.

16 Jonathan Crowe et al., 'Understanding the Legal Information Experience of Non-Lawyers: Lessons from the Family Law Context' (2018) 27(4) *Journal of Judicial Administration* 137, 141.

17 Felicity Bell, 'Family Law, Access to Justice, and Automation' (2019) 19 *Macquarie Law Journal* 103, 115.

18 Tania Sourdin, 'Justice and Technological Innovation' (2015) 25 *Journal of Judicial Administration* 96, 105.

19 OurDivorceAgreement, *OurDivorceAgreement* <http://www.ourdivorceagreement.com/>.

20 Settify, Settify <https://www.settify.com.au/>.

21 Legal Services Commission of South Australia, *LegalAidSA* (2019) <https://play.google.com/store/apps/details?id=com.andromo.dev86688.app95890>.

22 Women's Legal Service Queensland, *PENDA* (2017) <https://penda-app.com>.

23 Belinda Smyth and Bruce Fehlberg, 'Australian Post-Separation Parenting on the Smartphone: What's 'App-ening?' (2019) 41(1) *Journal of Social Welfare and Family Law* 53, 56.

24 Our Family Wizard, *Our Family Wizard* <https://www.ourfamilywizard.com.au/>.

25 Our Family Wizard, *Our Family Wizard* <https://www.ourfamilywizard.com.au/>.

26 MyMob, *MyMob* <https://www.mymob.com/>.

27 Amicable, *The Amicable Divorce App* <https://amicable.io/amicable-divorce-app/>.

28 Amicable, *The Amicable Divorce App* <https://amicable.io/amicable-divorce-app/>.

29 *My Divorce Notes* <https://apps.apple.com/sr/app/my-divorce-notes/id1163881321>.

30 Belinda Smyth and Bruce Fehlberg, 'Australian Post-Separation Parenting on the Smartphone: What's 'App-ening?' (2019) 41(1) *Journal of Social Welfare and Family Law* 53, 56.

31 Belinda Smyth and Bruce Fehlberg, 'Australian Post-Separation Parenting on the Smartphone: What's 'App-ening?' (2019) 41(1) *Journal of Social Welfare and Family Law* 53, 56.

32 Tania Sourdin, 'Justice and Technological Innovation' (2015) 25 *Journal of Judicial Administration* 96, 105.

33 See Tania Sourdin, 'Judge v Robot: Artificial Intelligence and Judicial Decision making' (2018) 41(4) *University of New South Wales Law Journal* 1114, 1131 nn 112.

34 See Emilia Bellucci and John Zeleznikow, 'How Online Negotiation Support Systems Empower People to Engage in Mediation: The Provision of Important Trade-off Advice' (2018) 5(1–2) *International Journal of Online Dispute Resolution* 94.

35 John Zeleznikow, 'Can Artificial Intelligence and Online Dispute Resolution Enhance Efficiency and Effectiveness in Courts' (2017) 8(2) *International Journal for Court Administration* 30, 41.

36 See Jin Ho Verdonschot, 'In the Netherlands, Online Application Helps Divorcing Couples in Their Own Words, on Their Own Times' (2015) 21(2) *Dispute Resolution Magazine* 19; Michael Legg, 'The Future of Dispute Resolution: Online ADR and Online Courts' (2016) 27(4) *Australasian Dispute Resolution Journal* 227, 230.

37 Peter Kenneth Cashman and Eliza Ginnivan, 'Digital Justice: Online Resolution of Minor Civil Disputes and the Use of Digital Technology in Complex Litigation and Class Actions' (2019) 19 *Macquarie Law Journal* 39, 46.

38 Maurits Barendrecht, 'Rechtwijzer: Why Online Supported Dispute Resolution is Hard to Implement', *Law Technology and Access to Justice* (Blog Post, 20 June 2017) <https://law-tech-a2j.org/odr/rechtwijzer-why-online-supported-dispute-resolution-is-hard-to-implement/>.

39 Belinda Smyth and Bruce Fehlberg, 'Australian Post-Separation Parenting on the Smartphone: What's 'App-ening?' (2019) 41(1) *Journal of Social Welfare and Family Law* 53, 56.

40 Belinda Smyth and Bruce Fehlberg, 'Australian Post-Separation Parenting on the Smartphone: What's 'App-ening?' (2019) 41(1) *Journal of Social Welfare and Family Law* 53, 65.

41 See Tania Sourdin, *Adieu Intelligent Divorce App and Family Dispute Resolution Project* (Final Report, 2020).

42 See 'Adieu', *Adieu: Elegant Parting* (Web Page, 2020) <https://www.adieu.ai/>.

43 See 'Lumi', *Adieu: Elegant Parting* (Web Page, 2020) <https://www.adieu.ai/lumi/>.

44 Tania Sourdin, 'Tech On Trial to Help Court System Access', *Newcastle Herald* (News Article, 31 July 2019) <https://www.newcastleherald.com.au/story/6302949/tech-on-trial-to-help-court-system-access/>.

45 'About Amica', *Amica* (Web Page, 2020) <https://amica.gov.au/about-amica.html>.

46 See Tania Sourdin, Bin Li, Stephanie Simms and Alexander Connolly, 'COVID-19, Technology and Family Dispute Resolution', (2020) *Australasian Dispute Resolution Journal* (Forthcoming); See also Shoshana Wodinsky, *Australian Authorities Want an AI To Settle Your Divorce*, GIZMODO, 1 July 2020 <https://www.gizmodo.com.au/2020/07/australian-authorities-want-an-ai-to-settle-your-divorce/>.

47 Felicity Bell, 'Family Law, Access to Justice, and Automation' (2019) 19 *Macquarie Law Journal* 103, 113.

48 Melissa Conley Tyler and Mark McPherson, 'Online Dispute Resolution and Family Disputes' (2006) 12(2) *Journal of Family Studies* 165, 170; Tania Sourdin and Chinthaka Liyanage, 'The Promise and Reality of Online Dispute Resolution in Australia' in Mohamed S Abdel Wahab, Ethan Katsh and Daniel Rainey (eds), *Online Dispute Resolution: Theory and Practice a Treatise on Technology and Dispute Resolution* (Eleven International Publishing, 2012) 483, 499.

49 Tania Sourdin and Chinthaka Liyanage, 'The Promise and Reality of Online Dispute Resolution in Australia' in Mohamed S Abdel Wahab, Ethan Katsh and Daniel Rainey (eds), *Online Dispute Resolution: Theory and Practice a Treatise on Technology and Dispute Resolution* (Eleven International Publishing, 2012) 483, 499.

50 Felicity Bell, 'Family Law, Access to Justice, and Automation' (2019) 19 *Macquarie Law Journal* 103, 119.

51 Emilia Bellucci, Sitalakshmi Venkatraman and Andrew Stranieri, 'Online Dispute Resolution in Mediating EHR Disputes: A Case Study on the Impact of Emotional Intelligence' (2019) *Behaviour & Information Technology* 3, 6.

52 Robert J Condlin, 'Online Dispute Resolution: Stinky, Repugnant, or Drab' (2017) 18(3) *Cardozo Journal of Conflict Resolution* 717, 751–2.

53 Belinda Smyth and Bruce Fehlberg, 'Australian Post-Separation Parenting on the Smartphone: What's 'App-ening?' (2019) 41(1) *Journal of Social Welfare and Family Law* 53, 62.

54 Belinda Smyth and Bruce Fehlberg, 'Australian Post-Separation Parenting on the Smartphone: What's 'App-ening?' (2019) 41(1) *Journal of Social Welfare and Family Law* 53, 60.

55 Suzie Forell, Meg Laufer and Erol Digiusto, 'Legal Assistance by Video Conferencing: What Is Known?' (Justice Issues Paper 15, Law and Justice Foundation of New South Wales, November 2011) 3.

56 Suzie Forell, Meg Laufer and Erol Digiusto, 'Legal Assistance by Video Conferencing: What Is Known?' (Justice Issues Paper 15, Law and Justice Foundation of New South Wales, November 2011) 1.

57 Tania Sourdin and John Zeleznikow, 'Courts, Mediation and COVID-19', *Australian Business Law Review* 48 (2020) 138.

58 See for example <https://law-tech-a2j.org/odr/justice-must-not-trump-efficiency-remote-courts-covid-19-and-the-justice-committee/>.

59 Suzie Forell, Meg Laufer and Erol Digiusto, 'Legal Assistance by Video Conferencing: What Is Known?' (Justice Issues Paper 15, Law and Justice Foundation of New South Wales, November 2011) 3.

60 John Zeleznikow, 'Using Artificial Intelligence to Support to Provide User Centric Intelligent Negotiation Support' (2020) 29 *Group Decision and Negotiation* (submitted).

61 John Zeleznikow, 'Using Artificial Intelligence to Support to Provide User Centric Intelligent Negotiation Support' (2020) 29 *Group Decision and Negotiation* (submitted). Zeleznikow has noted that such advisory tools may include books, articles, cases, legislation and videos; there would also be calculators.

62 John Zeleznikow, 'Using Artificial Intelligence to Support to Provide User Centric Intelligent Negotiation Support' (2020) 29 *Group Decision and Negotiation* (submitted). Zeleznikow explains that such tools are to enable negotiation, mediation, conciliation or facilitation of matters.

63 John Zeleznikow, 'Using Artificial Intelligence to Support to Provide User Centric Intelligent Negotiation Support' (2020) 29 *Group Decision and Negotiation* (submitted). Zeleznikow submits that if the disputants cannot resolve their conflict, software using game theory or artificial intelligence can be used to facilitate trade-offs.

64 John Zeleznikow, 'Using Artificial Intelligence to Support to Provide User Centric Intelligent Negotiation Support' (2020) 29 *Group Decision and Negotiation* (submitted). Zeleznikow explains that if and once a negotiation settlement is reached, software can be used to draft suitable agreements.

65 See generally 'What Is Immediation?', *Immediation* (Web Page) <https://www.immediation.com/>.

66 See generally 'Resolve the World's Disputes. Whenever. Wherever', *MODRON* (Web Page) <https://www.modron.com/>. MODRON is the provider favoured by the Australian Resolution Institute: 'Resolution Institute and MODRON Have Partnered to Bring Our Members Spaces', *Resolution Institute* <https://www.resolution.institute/resources/online-dispute-resolution-platforms/modron>.

67 See generally 'Better Co-Parenting, Happier Kids' *Our Family Wizard* (Web Page) <https://www.ourfamilywizard.com.au/>; Allan Barsky, 'The Ethics of App-Assisted Family Mediation' (2016) 34(1) *Conflict Resolution Quarterly* 31.

68 Tania Sourdin and John Zeleznikow, 'Courts, Mediation and COVID-19', *Australian Business Law Review* 48 (2020) 138.

69 Suzie Forell, Meg Laufer and Erol Digiusto, 'Legal Assistance by Video Conferencing: What Is Known?' (Justice Issues Paper 15, Law and Justice Foundation of New South Wales, November 2011) 10.

70 Suzie Forell, Meg Laufer and Erol Digiusto, 'Legal Assistance by Video Conferencing: What Is Known?' (Justice Issues Paper 15, Law and Justice Foundation of New South Wales, November 2011) 2.

71 Suzie Forell, Meg Laufer and Erol Digiusto, 'Legal Assistance by Video Conferencing: What Is Known?' (Justice Issues Paper 15, Law and Justice Foundation of New South Wales, November 2011) 3.

72 See generally Felicity Bell, 'Family Law, Access to Justice, and Automation' (2019) 19 *Macquarie Law Journal* 103, 119.

73 National Alternative Dispute Resolution Advisory Council, *Dispute Resolution and Information Technology: Principles for Good Practice* (Draft, March 2002) 12; Suzie Forell, Meg Laufer and Erol Digiusto, 'Legal Assistance by Video Conferencing: What Is Known?' (Justice Issues Paper 15, Law and Justice Foundation of New South Wales, November 2011) 2.

74 See Vivi Tan, 'Online Dispute Resolution for Small Civil Claims in Victoria: A New Paradigm in Civil Justice' (2019) 24(1) *Deakin Law Review* 101, 126.

75 National Alternative Dispute Resolution Advisory Council, *Dispute Resolution and Information Technology: Principles for Good Practice* (Draft, March 2002) 12; Suzie Forell, Meg Laufer and Erol Digiusto, 'Legal Assistance by Video Conferencing: What Is Known?' (Justice Issues Paper 15, Law and Justice Foundation of New South Wales, November 2011) 2.

76 Suzie Forell, Meg Laufer and Erol Digiusto, 'Legal Assistance by Video Conferencing: What Is Known?' (Justice Issues Paper 15, Law and Justice Foundation of New South Wales, November 2011) 11.

77 Suzie Forell, Meg Laufer and Erol Digiusto, 'Legal Assistance by Video Conferencing: What Is Known?' (Justice Issues Paper 15, Law and Justice Foundation of New South Wales, November 2011) 13.

78 Suzie Forell, Meg Laufer and Erol Digiusto, 'Legal Assistance by Video Conferencing: What Is Known?' (Justice Issues Paper 15, Law and Justice Foundation of New South Wales, November 2011) 13.

79 Suzie Forell, Meg Laufer and Erol Digiusto, 'Legal Assistance by Video Conferencing: What Is Known?' (Justice Issues Paper 15, Law and Justice Foundation of New South Wales, November 2011) 14.

80 Naom Ebner, 'Negotiation via Videoconferencing' in Chris Honeyman and Andrea Kupfer Schneider (eds), *The Negotiator's Desk Reference* (DRI Press, 2017) 151, 162.

81 Naom Ebner, 'E-Mediation' in Mohamed S Abdel Wahab, Ethan Katsh and Daniel Rainey (eds), *Online Dispute Resolution: Theory and Practice* (Eleven International Publishing, 2012) 357, 366.

82 Zoom 5.0 added a new encryption standard called AES 256-bit GCM encryption, which is said to be the gold standard of encryption and is the form used by the US government to secure data. On 23 April 2020, Zoom indicated that it would not be selling any user data after it was revealed that Zoom sends data from users of its IOS app to Facebook for advertising purposes, even if the user does not have a Facebook account. Hank Schless, a senior manager at tech security firm Lookout said: "The widespread use of conferencing solutions like zoom shows how people are OK with putting convenience ahead of security" See <https://www.theguardian.com/technology/2020/apr/23/zoom-update-security-encryption-bombing>.

83 Leanne Ho, 'Pro Bono Legal Services via Video Conferencing: Opportunities and Challenges' (Conference Paper, Australian Pro Bono Centre, 3rd National Rural Law and Justice Conference, 3–4 July 2015) 3.

84 Suzie Forell, Meg Laufer and Erol Digiusto, 'Legal Assistance by Video Conferencing: What Is Known?' (Justice Issues Paper 15, Law and Justice Foundation of New South Wales, November 2011) 14.

85 See the Guardian report available at <https://www.theguardian.com/uk-news/2020/apr/22/court-hearings-via-video-risk-unfairness-for-disabled-people?CMP=share_btn_tw> referring to an interim report by the UK Equality and Human Rights Commission (EHRC).

86 See report in the Conversation on 8 April 2020 referring to work by Rossner and David Tait available at <https://theconversation.com/courts-are-moving-to-video-during-coronavirus-but-research-shows-its-hard-to-get-a-fair-trial-remotely-134386?utm_source=twitter&utm_medium=bylinetwitterbutton>.

87 National Alternative Dispute Resolution Advisory Council, *Dispute Resolution and Information Technology: Principles for Good Practice* (Draft, March 2002) 12.

88 National Alternative Dispute Resolution Advisory Council, *Dispute Resolution and Information Technology: Principles for Good Practice* (Draft, March 2002) 10.

89 Vivi Tan, 'Online Dispute Resolution for Small Civil Claims in Victoria: A New Paradigm in Civil Justice' (2019) 24(1) *Deakin Law Review* 101, 127.

90 National Alternative Dispute Resolution Advisory Council, *Dispute Resolution and Information Technology: Principles for Good Practice* (Draft, March 2002) 10.

91 Noam Ebner and Jed Thompson, '@ Face Value? Nonverbal Communication & Trust Development in Online Video-Based Mediation' (2014) *International Journal of Online Dispute Resolution* <https://ssrn.com/abstract=2395857> 16.

92 Noam Ebner and Jed Thompson, '@ Face Value? Nonverbal Communication & Trust Development in Online Video-Based Mediation' (2014) *International Journal of Online Dispute Resolution* <https://ssrn.com/abstract=2395857> 6–7.

93 Naom Ebner, 'Negotiation via Videoconferencing' in Chris Honeyman and Andrea Kupfer Schneider (eds), *The Negotiator's Desk Reference* (DRI Press, 2017) 151, 162.

94 Robert J Condlin, 'Online Dispute Resolution: Stinky, Repugnant, or Drab' (Research Paper No 2016-40, University of Maryland, 27 December 2016) 30.

95 Leanne Ho, 'Pro Bono Legal Services via Video Conferencing: Opportunities and Challenges' (Conference Paper, Australian Pro Bono Centre, 3rd National Rural Law and Justice Conference, 3–4 July 2015) 3.

96 Robert J Condlin, 'Online Dispute Resolution: Stinky, Repugnant, or Drab' (Research Paper No 2016-40, University of Maryland, 27 December 2016) 31.

97 Suzie Forell, Meg Laufer and Erol Digiusto, 'Legal Assistance by Video Conferencing: What Is Known?' (Justice Issues Paper 15, Law and Justice Foundation of New South Wales, November 2011) 1.

98 Suzie Forell, Meg Laufer and Erol Digiusto, 'Legal Assistance by Video Conferencing: What Is Known?' (Justice Issues Paper 15, Law and Justice Foundation of New South Wales, November 2011) 2.

99 National Alternative Dispute Resolution Advisory Council, *Dispute Resolution and Information Technology: Principles for Good Practice* (Draft, March 2002) 11.

100 For a summary of arrangements see Tania Sourdin and John Zeleznikow, 'Courts, Mediation and COVID-19', *Australian Business Law Review* 48 (2020) 138.

101 Denise King, 'Internet Mediation – A Summary' (2000) 11(3) *Australian Dispute Resolution Journal* 180, cited in National Alternative Dispute Resolution Advisory Council, *Dispute Resolution and Information Technology: Principles for Good Practice* (Draft, March 2002) 10.

102 Susan Nauss Exon and Soomi Lee, 'Building Trust Online: The Realities of Telepresence for Mediators Engaged in Online Dispute Resolution' (2019) 49(1) *Stetson Law Review* (forthcoming).

103 Aaron Haas, 'Videoconferencing in Immigration Proceedings' (2006) 5(1) *Pierce Law Review* 59, 75.

104 Aaron Haas, 'Videoconferencing in Immigration Proceedings' (2006) 5(1) *Pierce Law Review* 59, 72.

105 Caroline Hart, "Better Justice?' or 'Shambolic Justice?': Governments' Use of Information Technology for Access to Law and Justice, and the Impact on Regional and Rural Legal Practitioners' (2017) 1 *International Journal of Rural Law and Policy* 1, 15. See also Amy Salyzyn, 'A New Lens: Reframing the Conversation about the Use of Video Conferencing in Civil Trials in Ontario' (2012) 50(2) *Osgoode Hall Law Journal* 429, 456.

106 Vivi Tan, 'Online Dispute Resolution for Small Civil Claims in Victoria: A New Paradigm in Civil Justice' (2019) 24(1) *Deakin Law Review* 101, 127.

107 Ron S Foster and Lianne M Cihlar, 'Technology and Family Law Hearings' (2014) 5(1) *Western Journal of Legal Studies* 1, 16.

108 Amy Salyzyn, 'A New Lens: Reframing the Conversation about the Use of Video Conferencing in Civil Trials in Ontario' (2012) 50(2) *Osgoode Hall Law Journal* 429, 447.

109 See for example concerns expressed relating to jury trials at <https://www.abajournal.com/web/article/could-zoom-jury-trials-become-a-reality-during-the-pandemic> and 'sentenced to death via zoom' at <https://edition.cnn.com/2020/05/07/africa/nigeria-zoom-death-sentence-intl/index.html>.

110 See Adam M. Gershowitz, 'Criminal-Justice Apps' (2019) 105 *Virginia Law Review Online*.

111 See the official website of Sydney Criminal Lawyers, <https://www.sydneycriminallawyers.com.au/media-centre/apps/nsw-pocket-lawyer/>.

112 See the official website of Go To Court Lawyers, <https://www.gotocourt.com.au/legal-news/legal-advice-app/>.

113 Anne Bowen Poulin, 'Criminal Justice and Videoconferencing Technology: The Remote Defendant' (2003–2004) 78 *Tulane Law Review* 1089, 1118.

114 Molly Treadway Johnson and Elizabeth C Wiggins, 'Videoconferencing in Criminal Proceedings: Legal and Empirical Issues and Directions for Research' (2006) 28(2) *Law & Policy* 211, 216.

115 Molly Treadway Johnson and Elizabeth C Wiggins, 'Videoconferencing in Criminal Proceedings: Legal and Empirical Issues and Directions for Research' (2006) 28(2) *Law & Policy* 211, 216.

116 Shari Seidman Diamond et al., 'Efficiency and Cost: The Impact of Videoconferenced Hearings on Bail Decisions' (2010) 100(3) *Journal of Criminal Law and Criminology* 869, 898–900.

117 Frank M Walsh and Edward M Walsh, 'Effective Processing or Assembly-Line Justice? The Use of Teleconferencing in Asylum Removal Hearings' (2008) 11 *Georgetown Immigration Law Journal* 259, 271.

118 See Meredith Rossner and Martha McCurdy, *Implementing Video Hearings (Party-to-State): A Process Evaluation* (2018) UK, Ministry for Justice, The London School of Economics and Political Science available at <http://eprints.lse.ac.uk/90960/>.

119 *State v. Loomis*, 371 Wis 2d, 235 (2016).

120 Cary Coglianese and Lavi Ben Dor, 'AI in Adjudication and Administration: A Status Report on Governmental Use of Algorithmic Tools in the U.S.' (2019) *University of Pennsylvania Law School, Public Law Research Paper* 19–41; Jordan Rodu and Michael Baiocchi, 'The Principled Prediction-Problem Ontology: When Black Box Algorithms Are (Not) Appropriate' (2020) *preprint arXiv:2001.07648*.

121 See Julia Angwin, Jeff Larson, Surya Mattu and Lauren Kirchner, 'Machine Bias', *ProPublica* <https://www.propublica.org/article/machine-bias-risk-assessments-in-criminal-sentencing>.

122 See Tania Sourdin, *Judges, Technology and AI*, forthcoming (2021) (Edward Elgar).

123 Julia Dressel and Henry Farid, 'The Accuracy, Fairness, and Limits of Predicting Recidivism' (2018) 4(1) *Science Advances*, <https://advances.sciencemag.org/content/4/1/eaao5580>.

124 Zhiyuan Lin et al., 'The Limits of Human Predictions of Recidivism' (2020) 6 *Science Advances* 1.

125 Zhiyuan Lin et al., 'In the US Criminal Justice System, Algorithms Help Officials Make Better Decisions, Our Research Finds', *Washington Post* (2 March 2020), <https://www.washingtonpost.com/politics/2020/03/02/us-criminal-justice-system-algorithms-do-help-officials-make-better-decisions-our-research-finds/>.

126 See Tania Sourdin, *Judges, Technology and AI*, forthcoming (2021) (Edward Elgar).

127 John Villasenor and Virginia Foggo, 'Algorithms and Sentencing: What Does Due Process Require?', *Brookings*, <https://www.brookings.edu/blog/techtank/2019/03/21/algorithms-and-sentencing-what-does-due-process-require/>.

128 John Villasenor and Virginia Foggo, 'Algorithms and Sentencing: What Does Due Process Require?', *Brookings*, <https://www.brookings.edu/blog/techtank/2019/03/21/algorithms-and-sentencing-what-does-due-process-require/>.

129 Electronic Privacy Information Center, 'Algorithms in the Criminal Justice System: Pre-Trial Risk Assessment Tools', <https://epic.org/algorithmic-transparency/crim-justice/>.

130 Pretrial Justice Institute, <https://www.pretrial.org/wp-content/uploads/Risk-Statement-PJI-2020.pdf>.

131 Judicial interpretation, regularly released by the SPC, is arguably one of sources of law in China. In this regard, refer to Supreme People's Court Monitor for more information: <https://supremepeoplescourt monitor.com/tag/judicial-interpretations/>.

132 Yadong Cui, '"Artificial Intelligence" Makes the Court System More Just, Efficient and Authoritative', <https://law.stanford.edu/china-law-and-policy-association-clpa/articles/>.

133 Yadong Cui, '"Artificial Intelligence" Makes the Court System More Just, Efficient and Authoritative', <https://law.stanford.edu/china-law-and-policy-association-clpa/articles/>.

134 See Weimin Zuo, 'Some Thoughts on Prospects of the Application of Legal Artificial Intelligence in China', (2018) 2 *Tsinghua University Law Journal*, 115–7.

135 See Chao Ma, Xiaohong Yu and Haibo He, 'Big Data Analysis: Open Report on China's Effort to Move Court Judgements Online' (2016) 4 *China Law Review* 195.

136 See Tania Sourdin, *Judges, Technology and AI*, forthcoming (2021) (Edward Elgar) and Tania Sourdin, 'Judge v Robot: Artificial Intelligence and Judicial Decision making' (2018) 41(4) *University of New South Wales Law Journal* 1114.

137 See, for example: Ithaca Group, *Evaluation of PENDA: A Financial Empowerment App for Women Escaping Domestic and Family Violence* (Final Report, June 2018) 16.

138 Roger Smith, 'Rechtwijzer: Why Online Supported Dispute Resolution Is Hard to Implement', *Law Technology and Access to Justice* (Blog Post, 20 June 2017) <https://law-tech-a2j.org/odr/rechtwijzer-why-online-supported-dispute-resolution-is-hard-to-implement/>.

139 Roger Smith, 'Rechtwijzer: Why Online Supported Dispute Resolution Is Hard to Implement', *Law Technology and Access to Justice* (Blog Post, 20 June 2017) <https://law-tech-a2j.org/odr/rechtwijzer-why-online-supported-dispute-resolution-is-hard-to-implement/>.

140 One participant commented that *Lumi* allowed them to take control of their divorce process post-agreement.

141 Roger Smith, 'Rechtwijzer: Why Online Supported Dispute Resolution Is Hard to Implement', *Law Technology and Access to Justice* (Blog Post, 20 June 2017) <https://law-tech-a2j.org/odr/rechtwijzer-why-online-supported-dispute-resolution-is-hard-to-implement/>.

142 Civil Resolution Tribunal, *Participant Satisfaction Survey – July 2019* (Web Page, 2019) <https://civilresolutionbc.ca/participant-satisfaction-survey-july-2019/>.

143 Lyria Bennett Moses, 'Artificial Intelligence in the Courts, Legal Academia and Legal Practice' (2017) 91(7) *Australian Law Journal* 561, 567–8.

144 Lyria Bennett Moses, 'Artificial Intelligence in the Courts, Legal Academia and Legal Practice' (2017) 91(7) *Australian Law Journal* 561, 567.

145 See for example <https://www.telegraph.co.uk/technology/2018/07/13/ai-doctor-app-babylon-fails-diagnose-heart-attack-complaint/>.

146 Dana Remus and Frank Levy, 'Can Robots Be Lawyers: Computers, Lawyers, and the Practice of Law' (2017) 30(3) *Georgetown Journal of Legal Ethics* 501.

147 Felicity Bell, 'Family Law, Access to Justice, and Automation' (2019) 19 *Macquarie Law Journal* 103, 131–2.
148 Felicity Bell, 'Family Law, Access to Justice, and Automation' (2019) 19 *Macquarie Law Journal* 103, 109.
149 Heather Scheiwe Kulp, 'Future Justice? Online Dispute Resolution and Access to Justice', *Just Court ADR* (Blog Post, 8 August 2011) <http://blog.aboutrsi.org/2011/policy/future-justice-online-dispute-resolution-and-access-to-justice/>.
150 Heather Scheiwe Kulp, 'Future Justice? Online Dispute Resolution and Access to Justice', *Just Court ADR* (Blog Post, 8 August 2011) <http://blog.aboutrsi.org/2011/policy/future-justice-online-dispute-resolution-and-access-to-justice/>.
151 Belinda Smyth and Bruce Fehlberg, 'Australian Post-Separation Parenting on the Smartphone: What's 'App-ening?' (2019) 41(1) *Journal of Social Welfare and Family Law* 53, 62.
152 Belinda Smyth and Bruce Fehlberg, 'Australian Post-Separation Parenting on the Smartphone: What's 'App-ening?' (2019) 41(1) *Journal of Social Welfare and Family Law* 53, 60.
153 Julie Macfarlane, *The National Self-Represented Litigants Project: Identifying and Meeting the Needs of Self-Represented Litigants* (Final Report, May 2013) 67.
154 See, eg, Samuel V Schoonmaker IV, 'Withstanding Disruptive Innovation: How Attorneys Will Adapt and Survive Impending Challenges from Automation and Nontraditional Legal Services Providers' (2017) 51(2–3) *Family Law Quarterly* 133.
155 Bregje Dijksterhuis, 'The Online Divorce Resolution Tool Rechtwijzer uit Elkaar Examined' in Mavis Maclean and Bregje Dijksterhuis (eds), *Digital Family Justice: From Alternative Dispute Resolution to Online Dispute Resolution?* (Hart Publishing, 2019).
156 Jeff Giddings and Michael Robertson, 'Informed Litigants with Nowhere to Go': Self-help Legal Aid Services in Australia' (2001) 26(4) *Alternative Law Journal* 184, 188.
157 Jeff Giddings and Michael Robertson, 'Informed Litigants with Nowhere to Go': Self-help Legal Aid Services in Australia' (2001) 26(4) *Alternative Law Journal* 184, 188.
158 Belinda Smyth and Bruce Fehlberg, 'Australian Post-Separation Parenting on the Smartphone: What's 'App-ening?' (2019) 41(1) *Journal of Social Welfare and Family Law* 53, 62.

5 Issues with justice apps

Introduction

As noted previously, apps in the justice sector can vary significantly. There are some apps that are neither designed for the justice sector nor necessarily repurposed or redesigned when used in the justice area. Examples include apps that support the exchange of documentation as well as videoconferencing, such as web-based platforms that include *Teams, Skype, Zoom, Google Hangouts,* and *WebEx* (general usage apps). Other apps are more specifically designed and directed at justice professionals such as lawyers (e.g., *Ross Intelligence* which is designed to assist lawyers with online research).[1] There are also apps that are directed at litigants, offenders, and others who work in the justice sector (e.g., police and corrective services staff) and others that are clearly designed to support court and tribunal operations (with several online handling platforms in China and the CRT in Canada).

The vast variation in terms of justice apps necessarily means that the various issues and limitations associated with app usage in the justice sector may depend not only on the types of technology used (e.g., supportive, replacement or disruptive, or some combination) but also the purpose and audience of the app. Some key issues that can arise with justice app use are discussed in this chapter and include: (i) the existence of a digital divide and accessibility issues, (ii) the challenges associated with translating law into code, (iii) the discretionary nature of legal decision-making, (iv) justice considerations, and (v) privacy and security concerns.

Some issues will be more relevant to some types of apps than others. For example, while access to justice issues, justice considerations and privacy and security concerns will be relevant when assessing generalist apps that are used in the justice sector, other issues will be less relevant. In addition, in some countries some matters may be more relevant than others because of domestic legal arrangements that may provide some additional protections and support. For example, privacy and security concerns may differ in EU countries as a result of broader legal requirements and where there is 'a right to explanation'[2] in respect of apps that can make automated decisions. There may also be differences between countries that have some common legal frameworks (as in EU countries) where such a right may be interpreted differently.

The digital divide and accessibility

As Internet use continues to expand across the globe, it is often assumed that people will be able to access ODR options and justice apps via the internet. However, the existence of a 'digital divide' means this is not always the case.[3] 'Digital divide' issues can impact upon the use and utility of all forms of new technology, and a range of factors contribute to this phenomenon around the globe. Whilst the digital divide may impact all people in terms of the capacity to use any apps or be informed or knowledgeable about a range of issues, in the justice sector there are particular issues that arise in terms of access to justice, particularly where there is an assumption that people can use an app and access the technology required to do so.

As such, how the digital divide may affect the delivery of 'justice' should be carefully considered, particularly given that promoting 'justice' should be one of the core purposes of technological innovation in law.[4] The issue of digital exclusion and its resultant social inequality have been highlighted in the context of the COVID-19 pandemic as more justice services are pushed online, leaving vulnerable groups in a disadvantaged position.[5]

Cabral et al. explain that the digital divide 'institutionalizes a two-tiered system incapable of delivering appropriate justice to low-income persons'.[6] As outlined by Toohey et al, however, it is not just individuals from lower socioeconomic communities who may have difficulty accessing digital services.[7] Elderly people, individuals with disabilities, indigenous people, and those who speak English as a second language can also face particular difficulties both using and accessing technology.[8] Sourdin, Li and Burke, have similarly identified an 'uneven' readiness to adopt new technologies that can be linked to geographical location, age, economic circumstances, and other factors related to vulnerability.[9]

One writer has optimistically referred to 'digital inclusion' issues.[10] These can be grouped into a number of areas that can affect how people access the internet and supportive technologies:

- culture and preference;
- broadband issues;
- age;
- disability;
- income;
- geographical factors; and
- education.

Importantly, some groups affected by these issues, including elderly people, individuals with disabilities, indigenous people, people who speak a language other than the common language in their country and people from lower socioeconomic communities, often have significant legal needs but also often face difficulties in accessing digital services. Such difficulties may also be related to their ability to access high bandwidth services and capacity to purchase a smartphone or computer, or access a device when required. Some of these issues relate to global inequalities and others relate to inequalities within countries. For example, in this regard, there remain significant global differences amongst populations in terms of internet usage during the COVID-19 pandemic.[11]

Some research from the UK that considers the use of remote hearings raises issues about whether some more vulnerable people might be excluded from the justice system or face difficulties where technology is used to enable remote access. For example, there have been two recent reports in the UK that have considered the impact of changes made as a result of the COVID-19 pandemic and have more specifically considered the use of video conferencing. The first report, by the Civil Justice Council (CJC) was intended to: 'understand the impact of the arrangements necessitated by COVID-19 on court users; make practical recommendations to address any issues over the short to medium term; and inform thinking about a longer-term review'.[12]

The CJC found that litigants' ability to initiate their case in person was suppressed, as was the capacity of vulnerable people to participate in court hearings during this time. As a result, the type of cases proceeding through the courts were fundamentally impacted.[13] For example, when lawyer respondents were asked about the most recent remote hearing they had participated in (480 hearings in total), data revealed that 47.3% of remote hearings had a monetary value of $50,000 (UK) or above.[14] Only 2.9 percent related to housing matters and 1.2 percent pertained to debt.[15] In addition, of the few hearings involving a litigant in person (10.9 percent),[16] over half were in relation to cases with a monetary value of less than $10,000 (UK).[17] In this regard, the CJC noted that:[18]

These findings would indicate that the proportion of vulnerable people and litigants in person participating in remote hearings may be artificially repressed by COVID-19 measures, with implications for findings regarding the efficacy of remote hearings.

Digital literacy issues also pose a challenge when considering the accessibility of ODR and justice apps. Here, questions arise as to whether older or less educated individuals are comfortable engaging with technology, or whether technologically enhanced dispute resolution is best suited to younger demographics. Giddings and Robertson have reported that there can be cultural issues and expectations surrounding the use of lawyers, with older people less likely to make use of self-help options.[19]

Cultural issues and language barriers can also give rise to accessibility issues. A 2008 report prepared for the Law Foundation of Ontario found that 'internet and other text-based solutions are of limited use to people who do not have the literacy skills to use them or to use them effectively'.[20] The report cautioned that 'vulnerable people, because they face language barriers, isolation, poverty, or a cluster of other difficulties that often accompany a legal problem, [ideally] need to receive direct services rather than to rely on self-help [through either digital or paper-based resources]'.[21]

Bailey, Burkell, and Reynolds have similarly noted that in the case of technological tools aimed at enhancing access to justice, there is a need to tailor the design of such tools to ensure they do not in fact exacerbate the access to justice gap for intended beneficiaries.[22] Such design strategies may include the use of plain language, the availability of content in multiple language formats, and design features to accommodate visual and other physical impairments.[23] An evaluation of the *PENDA* app created by the Women's Legal Service Queensland in Australia found that it needed to make greater use of icons and plain English in order to be accessible to those with intellectual disabilities.[24]

A related concern is present in relation to language. In many countries language remains a barrier. There may, for example, be situations where there is both a 'common' language and a 'formal' language and digital divide issues can replicate and possibly magnify issues that were already present in terms of access to the justice system. Whilst some apps may support more equitable arrangements by enabling more effective translation tools as well as language to text conversion, in some circumstances, app design may mean that inclusive arrangements are not fostered.

At the same time, it has been observed that the digital divide is no longer as significant an issue as it previously was. Sourdin, Li and Burke have noted that the digital divide has decreased as simpler technologies have evolved, Internet access has increased across communities, and technological competencies and preferences have grown.[25] Further, Bilinsky has noted that digital divide issues may vary according to the type of dispute or legal matter. That is, in a commercial dispute, digital divide issues may not, for example, be as significant, or may have little or no impact, yet such issues may be much more relevant in a civil dispute.[26]

In countries such as Australia, digital divide issues may be less relevant in family law disputes as many family, divorce, access, or support issues take place in families where the parties are under 35 years of age. This is significant because this demographic group is often familiar with, and has access to, technology.[27] On the other hand, Sourdin's preliminary findings in relation to the divorce app, *Adieu*, showed that the proportion of users of the app that had a relationship length of more than 20 years was more than 50 percent. This finding suggests that age may no longer be as relevant a factor in the context of the digital divide in some countries.[28]

There is another concern about the digital divide and the development of justice apps. This concern is that many justice apps are likely to be 'tested' on the most vulnerable members of the community or those that have little opportunity to complain when the app results in unfairness. Such apps may also lack 'explainability' or be 'opaque' as noted in relation to the *COMPAS* discussion (see Chapter 4). Such apps may include those developed for government administrative decision-making purposes (such as visa and social security debt purposes).

In this regard, there is also a concern that justice apps that are advisory, that is, justice apps which indicate the potential outcome of a court action, could result in a 'two-tiered' justice system, as those who cannot afford 'real' lawyers are forced to 'make do' with automated options.[29] There is some evidence of this in the medical setting where there can either be a reluctance to use a health app at all[30] (because of distrust or other issues), or where a health app is used for advisory purposes and the diagnosis is incorrect and/or further follow up or health care is not sought.[31]

Questions relating to the cost of advisory apps in the justice sector are also relevant (see further discussion in Chapter 6). Some apps may be publicly funded, funded through some philanthropic activity, be 'free', be costed via a download model, operate via subscription or as an 'add on' service, or may operate so that data is sold or used by a third party. The costing arrangements may raise particular issues for the most vulnerable members of a community and some people may not be able to afford automated advisory app options, highlighting the risk that a segment of the community may not be able to afford either a 'real' lawyer or an automated alternative.

Translating law into code

Some issues that arise with justice apps that are 'advisory' relate to the capacity of such apps to provide advice, or in some cases, make decisions based on understandings about the law. To some extent, these issues relate to the extent to which an app operates based on a designed legal expert system and/or is based on a machine learning platform. In relation to such systems, there are issues about what data is used to inform the decision-making (see Chapter 3 of this book), as well as questions about how law is interpreted and explained. Such issues may be more relevant in some justice systems depending on how laws are designed and created and how judges interpret the law.[32]

There are, however, moves to make it 'easier' for advisory systems to operate in the justice sector. For example, in December 2019, Australia's Commonwealth Scientific and Industrial Research Organisation ('CSIRO'), suggested to the Senate that Commonwealth legislation should not only be published in words but also in machine-readable code, allowing it to be read not only by lawyers but also computers.[33] In its submission, CSIRO argued that the move would boost the adoption of new regulatory technology across the economy, improving compliance while reducing costs. Essentially, CSIRO envisaged a future where the law could be directly applied by machines.

However, researchers have already highlighted the challenges associated with translating law into code. Justice Perry of the Federal Court of Australia has summarised some of the concerns:

> Computer programmers effectively assume responsibility for building decision making systems that translate policy and law into code. Yet computer programmers are not policy experts and seldom have legal training. How can we be sure that complex, even labyrinthal, regulations are accurately transposed into binary code?... We must be cautious of the human tendency to trust the reliability of computers.[34]

Whilst such cautions may be expressed about automated decision-making systems, concerns may not be as evident where advisory justice system apps are considered. In this regard, Susskind has noted that sometimes the law can be directly translated into code and referred to the *Divorce (Scotland) Act 1976* as an example of legislation faithfully coded into a programming language.[35] He has further noted that compared with legislation and statutory instruments, case law is harder to turn into computing language due to the difficulty of writing *ratio decidendi* with 'some simple canonical rule'.[36] Indeed, the role that judges may play in many countries in terms of the creation of law is significant.[37] As Justice Kirby in Australia has noted, legal innovation is not the top priority of parliaments, and correspondingly, the ability of judges to 'advance the law should be maintained'.[38] There is 'a need now more than ever for judges to fill in the legislative vacuum'.[39]

Ensuring accuracy in the substantive law applied by automated processes is challenging for a number of reasons, including, as noted by Sourdin, the fact the law in many countries operates within the context of statutory presumptions and discretionary judgments.[40] Justice Perry has similarly argued that the operation of statutory presumptions, and the fact that meaning is affected by context, means the potential for coding errors or distortions of meaning is real.[41]

Further, Justice Perry has noted that shades of meaning may be lost or distorted in any process of translation, and this poses a particular problem given our increasingly culturally diverse society. It is also noted that laws are not static, meaning automated systems will need to be capable of applying the law as it stands at previous points in time for decisions caught by transitional arrangements.[42] As summarised by Gershowitz, 'law is a complex beast that cannot be easily distilled into a simple questionnaire on a smartphone application'.[43] As noted previously, however, the different legal systems that operate around the world can mean that such factors are less relevant in some jurisdictions.

Despite this concern, a distinction can be made between 'advisory' apps – that is, those that advise people in the justice system what a probable outcome may be – and 'determinative' apps – that is, apps that result in automated decision-making. In relation to advisory justice apps, there are, as with advisory health apps, concerns that the 'answer' may not be correct and that it may be relied upon by a justice agency or individual to their detriment. It is partly for this reason that the design of justice apps that are advisory requires careful evaluation as well as testing (see Chapter 3). Indeed, McGill, Bouclin, and Salyzyn have noted that serious consideration must be given to whether a licensed lawyer should be involved in the creation and/or maintenance of justice apps to ensure the ongoing reliability of any legal information provided[44] (see discussion in Chapter 6 of this book relating to the unbundling of legal services).

In terms of determinative apps, a number of commentators have asserted that it is crucial that lawyers remain involved where fully automated technologies are used to make decisions.[45] According to Justice Perry, 'proper verification and audit mechanisms need to be integrated into the systems from the outset, and appropriate mechanisms for review in the individual case by humans, put in place'.[46] Such mechanisms can enhance the credibility of an app and can also ensure that outcomes that are reached are perceived to be 'just'. As summarised by Feldman, if ordinary citizens are to have faith in the credibility of artificial intelligence (AI), there must be methods of analysing and validating the choices made.[47] These observations are supported by an evaluation of *Rechtwijzer* in the Netherlands, discussed above, which found that although participants were satisfied with their experiences using the programme, a majority still felt the need to have a third party check over the agreement made through the system.[48]

The Honourable Tom Bathurst, Chief Justice of the Supreme Court of NSW in Australia, has raised a further issue: the fact that humans are irrational. This is problematic because 'systems which require definite inputs will inevitably fail to predict or answer human problems accurately'.[49] In this regard, not only must the law be clear (which is often not the case – or at least as perceived by one party in a dispute), but the inputs in terms of evidence may also be the subject of considerable disagreement. In relation to the *Adieu* app, and some other justice apps that are advisory rather than determinative, the variation in terms of understandings about the 'evidence' can be highlighted and flagged so that they are the subject of negotiation.

In the context of family law disputes, both Sourdin[50] (see also Chapter 3) and Bell have highlighted an additional issue with the use of automated decision-making systems (and potentially advisory systems) that rely on machine learning: the fact that any reliance on data comprised only of judgments would represent a collection of 'outlier' data, given that the majority of separations do not proceed to final hearing and judgment. In this regard, it is noted that the benefit of using a 'real' lawyer is their experience of settled as well as litigated cases.[51] The ALRC has similarly noted that in Australia, those litigating family law disputes represent only a very small proportion of all people who go through separation, with approximately 70 percent of people resolving parenting disputes without recourse to the family law system, and 40 percent resolving their property disputes via discussion. It is projected that this rate is higher for separating couples without children.[52] Of matters which do enter the system, the 'vast majority' settle.[53]

The issues with the data sample, as well as the supervision and review of justice apps that make decisions or provide advice, are somewhat different, although in relation to each there remain issues about how and to what extent law, which can be highly contextual, can be interpreted, even where simplified. Using a simplistic analogy from the health app area, it would be as though in diagnosing someone with a fracture to their arm, the fact that they were missing their other hand was not considered. That is, in the justice app area there is a concern that advice or actual decisions will be made in the absence of human input and that this will lead to unjust or unfair outcomes. The extent to which this is a valid concern may depend to some extent on the nature of the existing legal and court system. In addition, advisory justice apps are generally likely to be perceived in a more favourable way compared with determinative apps, particularly where some level of human review or oversight is maintained.

Discretion

Issues also exist around the exercise of discretion when considering the extent to which any judicial decision-making should be replaced by AI or a justice app.[54] Whilst such extensions are only developmental at this stage, there are many examples of automated administrative decision-making taking place. Again, such issues must be considered in the context of the judicial and court systems that operate in various countries around the world, as well as the subject matter of the dispute. For example, such issues may be particularly pertinent in the family law context. As outlined by Parkinson, in Australia, family law decisions are highly discretionary and there are no principles of quantification which can guide the resolution of property disputes.[55] Condlin has also noted that 'software is logical, not reasonable, and legal judgments often require both qualities in equal measure'.[56]

Speaking in relation to automated decision-making in the context of administrative decisions in Australia, Justice Perry has observed:

> Automated decision making systems are grounded in logic and rules-based programs that apply rigid criteria to factual scenarios. Importantly, they respond to input information entered by a user in accordance with predetermined outcomes. By contrast, many administrative decisions require the exercise of a discretion or the making of an evaluative judgment. These are complex and subtle

questions incapable of being transcribed into rigid criteria or rules and, therefore, beyond the capacity of an automated system to determine. Different factors may need to be weighed against each other and may be finely balanced. If automated systems were used in cases of this kind, not only may there be a constructive failure to exercise the discretion; by their nature they apply pre-determined outcomes raising questions of pre-judgment or bias.[57]

In practice, there have been some issues in relation to executive decision-making rooted in automation. For example, in Australia, the Commonwealth Government has used the Australian Centrelink debt recovery scheme ('Robodebt' scheme) to calculate and claim alleged overpayments from social security recipients.[58] However, the formula used has failed to produce the legally correct result in a significant percentage of cases,[59] and, in late 2019, the Federal Court of Australia confirmed the unlawfulness of some administrative decisions made by the automated system.[60]

The implicit biases that may be perpetuated by AI have also led some to argue that discretion should not be entirely removed from the equation. As outlined by Cruz, implicit biases in AI formulas can skew results in a way that negatively impact minority individuals.[61] This is because the designers of AI programmes typically come from very similar backgrounds: they are usually highly educated, cisgender men – most of whom are Caucasian or Asian – and their preferences do not reflect the beliefs, experiences, and preferences of marginalised communities.[62] Many would suggest that such criticisms could also be levelled at judges and administrative decision-makers who may also have a range of biases and may be more likely to come from particular backgrounds.

In the Chinese context, Ji has argued that automated judgments deriving from AI raise questions about whether humans or machines are 'judges'. As such, there is a risk that the independence of 'judges' could be violated by the combined intentions of programmers, software engineers, information technology companies, and other entities, so long as they participated in the design of the automated decision-making process.[63] This issue, and the extent to which it is a concern, varies according to country and approaches to the relationship between the judiciary and the executive. Put simply, in some countries, a clear separation of powers does not exist[64] (or not to the same extent as in other countries) and there may be less concern about executive involvement in judicial decision-making. Nevertheless, even where the independence of the judiciary may be a less relevant factor, the questions relating to the design and set up of justice apps that purport to mimic human judges remain relevant.[65]

In light of the above, it has been suggested that human decision-makers must be kept 'in the loop', and automation which requires the application of strict criteria, rather than the exercise of discretion or a value-based judgment, should be the subject of careful scrutiny in order to guard against unfair or arbitrary decisions.[66] Zeleznikow has similarly warned against ODR being fully automated, indicating that such systems should aim to support decision making rather than take over this function.[67]

Zalnieriute, Bennett Moses, and Williams have also identified a risk that automation can compromise individual due process rights by undermining the ability of a party to challenge a decision affecting them.[68] It is important, therefore, to ensure automated processes do not prevent parties from accessing or assessing the information used to make the decision. This has also been emphasised by Sourdin, Li, and Burke, who have identified the risk that the processes used to reach an outcome, particularly by more 'disruptive' technologies involving developed AI, may be less visible to the parties. The authors note the transparency and natural justice issues that can arise as a result of this.[69]

When it comes to keeping human decision-makers 'in the loop', Re and Solow-Niederman have outlined two possible options.[70] First, 'human and AI judges might collaborate by operating in tandem at specified stages of the judicial process'. This could occur either by preserving a measure of human oversight and involvement at particular points or incorporating human oversight at the front-end or back-end of a legal decision. A second option is to 'apportion discrete types of judicial decision-making to human as opposed to mechanized actors'. The resulting separation could be based on either subject matter, or a more finely grained determination about the parts of a legal decision that raise particular justice concerns. The authors note that this division of labour preserves 'a traditional role for humans within systems of AI adjudication, even if that role introduces increased opportunities for bias, arbitrariness, error, and cost'.[71] There is of course a third and important option that relates to the extent to which humans with expertise (lawyers, judges and others) are involved in the app design process.

At the same time, however, it is recognised that there are 'pragmatic difficulties' which can hinder the attempt to divide human and AI tasks in a way that desirably preserves human discretion, including an inability to know the right balance of human and AI activity prior to experimentation.[72] As such, there is a risk that 'pursuit of human-AI collaboration ... could end up being more like the worst of both worlds than the best if the wrong policy tradeoffs are struck'.[73]

Justice considerations

The issues with justice apps also require a consideration of the meaning of 'justice' and the extent to which such apps support substantive justice outcomes, as well as the extent to which they may support procedurally just processes. 'Justice' is of itself a broad concept (and may have many sub-objectives relating to distributive, restorative, and other elements).[74] In respect of procedural justice, two very separate notions of what is critical in terms of the attainment of justice have emerged. On the one hand, procedural justice in the context of modern courts is linked to case management and ensuring that an accurate decision is reached after following appropriate procedures.[75] This conception is related to case management within courts and is more relevantly linked to adjudicatory processes. In this regard, Bentham and others have produced a rich discourse with some disagreements relating to the interplay between procedural and substantive justice.[76]

As Sourdin has noted, a number of procedural justice theorists have concentrated on procedural justice and procedural fairness from a social science and psychological perspective. This literature focusses not only on what is done, but also on how people perceive the justice experience. Being treated respectfully, understanding processes and 'being heard' are critical components. The work of Lind and Tyler,[77] and the procedural justice work of Thibaut and Walker, suggest that, if people believe that they have been treated fairly, they are more likely to accept a decision and outcome.[78] These forms of justice are increasingly being measured in both qualitative and quantitative studies and are linked to measuring subjective experiences in the context of whether and how people experience justice.[79]

These different perceptions of procedural justice highlight some issues in the context of discussion about adjudicatory and non-adjudicatory forms of dispute finalisation and are therefore relevant when considering the different types of apps that are present in the justice sector. For example, supportive and replacement justice apps may encourage and support procedural justice. Apps that provide evaluative and determinative outcomes may be focussed on substantive justice and also justice in terms of process (linked to visibility and transparency). In assessing whether an app that enables video conferencing is 'just', a focus on both the outcome experienced, as well as the extent to which a person considered they understood the process, were able to contribute, and perceived that they were dealt with in a dignified manner, will be important in determining whether justice objectives are met by a justice app.

Justice can also be considered in terms of its location. That is, justice systems (as noted in Chapter 1) incorporate systems outside of courts and include the informal processes that people may use to seek justice, as well as the formal and institutional arrangements that exist to regulate behaviour.[80] In this way, justice apps can have a potentially wide ambit. For example, they can include negotiation support tools as well as complaint escalation apps.

In addition, concerns about whether or not justice objectives have been met can arise at both a structural and individual level.[81] When it comes to app usage in particular, Barsky has noted the risk that people will assume the use of the same technology by both parties to a dispute will ensure the process is fair and neutral when, in reality, this may not be the case if one party is more comfortable using a particular form of technology than the other.[82] The Women's Legal Service in Queensland in Australia has also raised the particular justice issues faced by women in this context. It is noted that women often suffer disadvantage if they attempt to 'help themselves' through the family law system, with self-help services 'not of sufficient benefit to effectively overcome the difficulties facing such women'.[83]

There is also a risk that a focus on the cheap and quick resolution of disputes will come at the cost of a just outcome. Condlin has questioned whether 'the cheap and efficient processing of disputes is a capitulation to the conditions of modern society more than a superior system for administering justice'.[84] Further, it is noted that ODR systems may restrict the ability of parties to argue the substantive merits of their claims:

> Uncoupling disputes from their substantive merits can undermine the fairness of individual outcomes and, if widespread, threaten the legitimacy of dispute resolution systems themselves.[85]

Condlin has further noted that ODR systems often require parties to explain their claims in fixed or pre-defined parts. As a result, there is a risk that ODR systems may not capture all the dimensions of the claim, and parties may not be able to recover the entire claim's worth.[86]

Re and Solow-Niederman have drawn a distinction between 'codified justice' and 'equitable justice':

> Codified justice aspires to establish the total set of legally relevant variables in advance, while discounting other facts and circumstances discoverable in individualized proceedings. The basic goal of such standardization is to reduce space for human discretion in adjudication, thereby diminishing opportunities for arbitrariness, bias, and waste, while increasing efficiency, consistency, and transparency. In short, codified justice sees the vices of discretion, whereas equitable justice sees its virtues.[87]

Efficiency and uniformity are identified as the 'main strengths of AI adjudication' and the 'two hallmarks of codified justice'.[88] The authors have argued that the rise in codified justice and decline in equitable justice leads to 'alienation' and the risk that 'important aspects of social life [are left] without sufficient public participation and oversight'.[89] Further, it is argued that decreased human involvement, or full automation, may make the operation of law 'seem that much less interesting, relevant, and subject to the control and care of everyday people'.[90]

In light of these concerns, there are four suggested responses to the decline in equitable justice. One of these involves integrating a measure of equitable justice into AI adjudication by 'coding equity' into the AI adjudicator itself.[91] To avoid the risk of 'locking in a baseline definition of equity that is 'aligned with extant values', Re and Solow-Niederman suggest that it would be preferable to update this coding at regular intervals. Further, it is noted that 'a machine capable of dispensing "AI equity" could also mitigate the problem of datafication by being even more responsive than human judges when it comes to a case's subtle factual nuances or changes in social values'.[92] Nevertheless, it is ultimately concluded that:

> Coding for equity is not a straightforward fix, however, in either a technical or a normative sense. It is not clear whether it is even technologically possible to code for nuanced equitable correction.[93]

Privacy and security

In Chapter 3, it is noted that justice apps also raise important privacy and security challenges.[94] Although there is a dearth of research exploring the privacy and security issues inherent in the use of justice apps, there has been a considerable amount of research exploring these issues in relation to ODR and apps more generally. McGill, Bouclin, and Salyzyn have noted that research in relation to health apps can provide useful insights for thinking about some of the privacy and security concerns relevant to justice apps.[95]

One of the most significant issues which arises relates to the risk that data collected by apps may be vulnerable to collection and misuse by unauthorised third parties. McGill, Bouclin, and Salyzyn highlight the fact that people may not fully understand what happens to the data that they share, including whether the rights to this data are reserved to the company collecting it, and whether the data can be shared for profit. It is noted that such concerns are particularly acute in relation to justice apps, given the sensitive personal information – including financial details – that such apps may collect.[96] Privacy concerns can also arise where a third party provider operating in a different jurisdiction is used to store data online.[97] To some extent, as noted in Chapter 6, these concerns can also be linked to the identity of the app developer and their commercial or non-commercial interest in terms of app development. In addition, in a number of jurisdictions, as discussed previously, there can be more well-developed legislation that can support privacy and data use relating to apps.

In specifically considering justice apps that support or replace legal processes, Scassa et al. have identified a number of unique privacy concerns which make justice apps different from other apps: (i) user confusion as to whether solicitor-client privilege or lawyer confidentiality applies in the context of justice apps which provide legal information or help users create legal documents; (ii) the collection of especially sensitive information relating to legal problems; (iii) heightened interest on the part of third parties (e.g., an opposing party in a lawsuit) in requesting or compelling the disclosure of the personal information collected; and (iv) special concerns arising where the app engages with court data or a public registry.[98]

In relation to the first of these concerns, McGill, Bouclin, and Salyzyn have noted that there may be confusion surrounding the confidentiality implications of justice apps. Where an app only provides legal information, as opposed to legal advice – meaning a lawyer-client relationship is not created – users may not understand that the confidentiality protections inherent in the lawyer-client relationship do not apply in the justice app context.[99] In relation to the second concern – the collection of especially sensitive information – Cruz has also noted that a privacy breach in the legal context can have 'devastating consequences due to the sensitive nature of the information communicated'.[100] Cruz has noted that 'the stakes are particularly high for low-income and immigrant individuals who may not understand that they are sharing their private information by using an app or filling out an online survey'.[101]

The above privacy issues may dissuade individuals from using justice apps, although given the development of apps in other fields, it would seem that many people are prepared to forgo privacy in relation to apps which may already gather data about who we engage with, what we buy and what we do. However, privacy issues may be of more particular concern to some groups and it has been noted that generational divides can correlate with scepticism in technology.[102] There is also a risk that people with particular concerns about the protection of their sensitive information – such as those with mental health issues, or, in the family law context, victims of domestic abuse – may be disinclined to use technology to address their legal issues.[103]

In light of the privacy concerns associated with app usage, particularly in the legal context, Scassa et al have developed a 'Privacy Code of Practice' for direct-to-public justice apps. This includes advice for app developers that in circumstances where the app does not create a lawyer-client relationship, it should be made 'very clear' to app users 'that the information they are sharing does not receive any special protections and could potentially be used against them in a legal proceeding'. The Privacy Code of Practice also outlines a number of 'design choices' that can support privacy. These include limiting the collection of personal information, and carefully considering whether data that is collected for research purposes and/or sold can be de-identified. It is also noted that a justice app can only collect personal data that is necessary to achieve the purposes for which the data is collected. Finally, the Privacy Code of Practice contains guidelines on how long app developers can keep the personal information they have collected through the app, and their data security obligations.[104]

Conclusions

For justice apps to be effective, as noted in Chapter 3, a range of matters must be considered when apps are being designed, used, reviewed, and extended. Many of those matters will determine whether an app supports justice objectives and does not have unintended adverse consequences. Of critical importance is whether an app can promote user trust in the content offered by the automated systems.[105]

'User trust' has been identified as encompassing information security, data security, personal security, and system security.[106] The notion also encompasses trust in the process and outcomes provided (linked therefore to notions of substantive and procedural justice). This broader notion of trust will be more relevant when justice apps are engaged in furnishing more specific or tailored information or advice.

As Ebner and Zeleznikow have noted, if a system is going to give advice about trade-offs or optimising agreements, there must be sufficient user trust in the algorithms which underlie and generate this advice. Of course, and as noted in Chapter 3, users must also be convinced that the technology: (i) will not fail or freeze up; (ii) will be able to support their dispute; (iii) will be competent in performing as promised; (iv) will not involve time or costs beyond what the consumer envisions; and (v) will be user-friendly.[107]

At present, with justice app development at a somewhat early stage (compared, e.g., with health app development), it might be tempting to consider that the obstacles are so significant that justice app developments should not proceed. Weighed against this, however, are the significant advantages that well-designed justice apps may offer in terms of cost and convenience, as well as opportunities to support access to justice and produce outcomes that are perceived to be fair and are considered to meet procedural justice requirements (and may do so more effectively than existing arrangements). It is important that measures adopted to ensure that justice apps address issues user privacy and security concerns do not discourage app innovation that would otherwise be in the public interest.[108]

Notes

1 See the official weblink to *Ross Intelligence*, available <https://www.rossintelligence.com/>.
2 See Lilian Edwards and Michael Veale, 'Enslaving the Algorithm: From a 'Right to an Explanation' to a 'Right to Better Decisions'?' (2018) 16(3) *IEEE Security & Privacy* 46–54.
3 Felicity Bell, 'Family Law, Access to Justice, and Automation' (2019) 19 *Macquarie Law Journal* 103, 124; Jena McGill, Suzanne Bouclin and Amy Salyzyn, 'Mobile and Web-based Legal Apps: Opportunities, Risks and Information Gaps' (2017) 15 *Canadian Journal of Law and Technology* 229, 246.
4 Tania Sourdin, Bin Li and Tom Hinds, 'Humans and Justice Machines: Emergent Legal Technologies and Justice Apps' (2020) 156 *Precedent* 23.
5 See Centre for Social Impact, *Digital Inclusion and COVID-19.* The document is available at <https://www.csi.edu.au/media/uploads/csi-covid_factsheet_digitalinclusion.pdf>.
6 James E. Cabral et al., 'Using Technology to Enhance Access to Justice' (2012) 26(1) *Harvard Journal of Law & Technology* 241, 265.

7 Lisa Toohey et al., 'Meeting the Access to Civil Justice Challenge: Digital Inclusion, Algorithmic Justice, and Human-Centred Design' (2019) 19 *Macquarie Law Journal* 133, 145.

8 Jeff Giddings and Michael Robertson, 'Informed Litigants with Nowhere to Go': Self-help Legal Aid Services in Australia' (2001) 26(4) *Alternative Law Journal* 184, 188; Lisa Toohey et al., 'Meeting the Access to Civil Justice Challenge: Digital Inclusion, Algorithmic Justice, and Human-Centred Design' (2019) 19 *Macquarie Law Journal* 133, 145.

9 Tania Sourdin, Bin Li and Tony Burke, 'Just Quick and Cheap? Civil Dispute Resolution and Technology' (2019) 19 *Macquarie Law Journal* 17, 34.

10 See M Wahab, 'The Global Information Society and Online Dispute Resolution: A New Dawn for Dispute Resolution' (2004) 21(2) *Journal of International Arbitration* 14.

11 See for example the Pew Research Report of April 2020 on this topic available at <https://www.pewresearch.org/fact-tank/2020/04/02/8-charts-on-internet-use-around-the-world-as-countries-grapple-with-covid-19/>.

12 Natalie Byrom, Sarah Beardon and Abby Kenrick, Civil Justice Council, *The Impact of COVID-19 Measures on the Civil Justice System* (Report, May 2020) 5.

13 Natalie Byrom, Sarah Beardon and Abby Kenrick, Civil Justice Council, *The Impact of COVID-19 Measures on the Civil Justice System* (Report, May 2020) 7, 21.

14 Natalie Byrom, Sarah Beardon and Abby Kenrick, Civil Justice Council, *The Impact of COVID-19 Measures on the Civil Justice System* (Report, May 2020) 7–8, 29.

15 Natalie Byrom, Sarah Beardon and Abby Kenrick, Civil Justice Council, *The Impact of COVID-19 Measures on the Civil Justice System* (Report, May 2020) 8, 21.

16 'Litigant in person' being self-represented litigants.

17 Natalie Byrom, Sarah Beardon and Abby Kenrick, Civil Justice Council, *The Impact of COVID-19 Measures on the Civil Justice System* (Report, May 2020) 8, 29.

18 Natalie Byrom, Sarah Beardon and Abby Kenrick, Civil Justice Council, *The Impact of COVID-19 Measures on the Civil Justice System* (Report, May 2020) 21.

19 Jeff Giddings and Michael Robertson, 'Informed Litigants with Nowhere to Go': Self-help Legal Aid Services in Australia' (2001) 26(4) *Alternative Law Journal* 184, 188.

20 Karen Cohl and George Thomson, *Connecting across Language and Distance: Linguistic and Rural Access to Legal Information and Services* (Report, 2008) 35.

21 Karen Cohl and George Thomson, *Connecting across Language and Distance: Linguistic and Rural Access to Legal Information and Services* (Report, 2008) 52.

22 Jane Bailey, Jacquelyn Burkell and Graham Reynolds, 'Access to Justice for All: Towards an 'Expansive Vision' of Justice and Technology' (2013) 31 *Windsor YB Access Just* 181, 198.

23 Jane Bailey, Jacquelyn Burkell and Graham Reynolds, 'Access to Justice for All: Towards an 'Expansive Vision' of Justice and Technology' (2013) 31 *Windsor YB Access Just* 181, 198.

24 Ithaca Group, *Evaluation of PENDA: A Financial Empowerment App for Women Escaping Domestic and Family Violence* (Final Report, June 2018) 6.

25 Tania Sourdin, Bin Li and Tony Burke, 'Just Quick and Cheap? Civil Dispute Resolution and Technology' (2019) 19 *Macquarie Law Journal* 17, 36.

26 David Bilinsky, 'Report from the ODR Conference in Buenos Aires', *Slaw* (Web Page, 3 June 2010) <http://www.slaw.ca/2010/06/03/report-from-the-odr-conference-in-buenos-aires/>.

27 David Bilinsky, 'Report from the ODR Conference in Buenos Aires', *Slaw* (Web Page, 3 June 2010) <http://www.slaw.ca/2010/06/03/report-from-the-odr-conference-in-buenos-aires/>.

28 Tania Sourdin, Bin Li and Tom Hinds, 'Humans and Justice Machines: Emergent Legal Technologies and Justice Apps' (2020) 156 *Precedent* 23.

29 Felicity Bell, 'Family Law, Access to Justice, and Automation' (2019) 19 *Macquarie Law Journal* 103, 131. See also Lisa Toohey et al., 'Meeting the Access to Civil Justice Challenge: Digital Inclusion, Algorithmic Justice, and Human-Centred Design' (2019) 19 *Macquarie Law Journal* 133, 146.

30 See discussion at <https://hcldr.wordpress.com/2019/07/30/technology-access-in-healthcare-the-digital-divide/>.

31 See earlier discussion relating to the Babylon health app as well as specific concerns raised in 2020 https://edmontonjournal.com/opinion/columnists/opinion-albertas-virtual-health-care-app-plagued-with-problems/.

32 For a full discussion of these factors see Tania Sourdin, *Judges, Technology and AI*, forthcoming (2021) (Edward Elgar).

33 See CSIRO Submission 19/691, *Financial Technology and Regulatory Technology* (2019) <http://www.aph.gov.au>.

34 Justice Melissa Perry, 'iDecide: Administrative Decision making in the Digital World' (2017) 91 *Australian Law Journal* 29, 32.

35 Richard Susskind, *Online Courts and The Future of Justice* (Oxford University Press, 2020), 160.

36 Richard Susskind, *Online Courts and The Future of Justice* (Oxford University Press, 2020), 160.

37 See Tania Sourdin, *Judges, Technology and AI*, forthcoming (2021) (Edward Elgar).

38 Michael Kirby, 'Judging: Reflections on the Moment of Decision' in Ruth Sheard (ed) *A Matter of Judgment Judicial decision-making and judgment writing* (Lexis Nexis Butterworths, 2003) 43, 45. President's Report, Extra-Judicial Notes (1997) 16 *Australian Bar Review* 2, 9.

39 Michael Kirby, 'Judging: Reflections on the Moment of Decision' in Ruth Sheard (ed) *A Matter of Judgment Judicial decision-making and judgment writing* (Lexis Nexis Butterworths, 2003) 43, 45, cited in Tania Sourdin, *Judges, Technology and AI*, forthcoming (2021) (Edward Elgar).

40 Tania Sourdin, 'Judge v Robot: Artificial Intelligence and Judicial Decision making' (2018) 41(4) *University of New South Wales Law Journal* 1114, 1127.

41 Justice Melissa Perry, 'iDecide: Administrative Decision making in the Digital World' (2017) 91 *Australian Law Journal* 29, 32.

42 Justice Melissa Perry, 'iDecide: Administrative Decision making in the Digital World' (2017) 91 *Australian Law Journal* 29, 32.

43 Adam M Gershowitz, 'Criminal Justice Apps' (2019) 105 *Virginia Law Review.*

44 Jena McGill, Suzanne Bouclin and Amy Salyzyn, 'Mobile and Web-based Legal Apps: Opportunities, Risks and Information Gaps' (2017) 15 *Canadian Journal of Law and Technology* 229, 250.

45 Justice Melissa Perry, 'iDecide: Administrative Decision making in the Digital World' (2017) 91 *Australian Law Journal* 29, 32.

46 Justice Melissa Perry, 'iDecide: Administrative Decision making in the Digital World' (2017) 91 *Australian Law Journal* 29, 32.

47 Robin C Feldman, 'Technology Law: Artificial Intelligence: Trust and Distrust' (2019) 3(17) *The Judges' Book* 115, 117.

48 Esmee A Bickel, Maria Anna Jozefa van Dijk and Ellen Giebels, *Online Legal Advice and Conflict Support: A Dutch Experience* (Report, University of Twente, March 2015) 22, 31.

49 Chief Justice Tom Bathurst, 'iAdvocate v Rumpole: Who will Survive? An Analysis of Advocates' Ongoing Relevance in the Age of Technology' (2015 Australian Bar Association Conference, Boston, 9 July 2015).

50 Tania Sourdin, 'Judge v Robot: Artificial Intelligence and Judicial Decision Making' (2018) 41(4) *University of New South Wales Law Journal* 1114, 1117–8.

51 Felicity Bell, 'Family Law, Access to Justice, and Automation' (2019) 19 *Macquarie Law Journal* 103, 118.

52 Australian Law Reform Commission, 'Family Law for the Future: An Inquiry into the Family Law System – Final Report' (Report No 135, March 2019) 79.

53 Australian Law Reform Commission, *Family Law for the Future: An Inquiry into the Family Law System – Final Report* (Report No 135, March 2019) 80.

54 Tania Sourdin, 'Judge v Robot: Artificial Intelligence and Judicial Decision making' (2018) 41(4) *University of New South Wales Law Journal* 1114, 1133.

55 Patrick Parkinson, 'Why Are Decisions on Family Property So Inconsistent?' (2016) 90(7) *Australian Law Journal* 498, 498–9.

56 Robert J Condlin, 'Online Dispute Resolution: Stinky, Repugnant, or Drab' (2017) 18(3) *Cardozo Journal of Conflict Resolution* 717, 723.

57 Justice Melissa Perry, 'iDecide: Administrative Decision Making in the Digital World' (2017) 91 *Australian Law Journal* 29, 33.

58 Monika Zalnieriute, Lyria Bennett Moses and George Williams, 'The Rule of Law and Automation of Government Decision Making' (2019) 82(3) *Modern Law Review* 425.

59 Monika Zalnieriute, Lyria Bennett Moses and George Williams, 'The Rule of Law and Automation of Government Decision-Making' (2019) 82(3) *Modern Law Review* 425, 446.

60 Victoria Legal Aid, *Explainer – Deanna Amato's robo-debt case*, <https://www.legalaid.vic.gov.au/about-us/news/explainer-deanna-amatos-robo-debt-case>.

61 Sherley Cruz, 'Coding for Cultural Competency: Expanding Access to Justice with Technology' (2019) 86 *Tennessee Law Review* 347, 369–71.

62 Sherley Cruz, 'Coding for Cultural Competency: Expanding Access to Justice with Technology' (2019) 86 *Tennessee Law Review* 347, 369–71.

63 See Weidong Ji, The Change of Jurisdiction in the Era of Artificial Intelligence, *Oriental Law* (2018), Issue 1, 131–2.

64 See Tania Sourdin, *Judges, Technology and AI*, forthcoming (2021) (Edward Elgar).

65 See Tania Sourdin, *Judges, Technology and AI*, forthcoming (2021) (Edward Elgar).

66 Justice Melissa Perry, 'iDecide: Administrative Decision making in the Digital World' (2017) 91 *Australian Law Journal* 29, 33–4.

67 John Zeleznikow, 'Can Artificial Intelligence and Online Dispute Resolution Enhance Efficiency and Effectiveness in Courts' (2017) 8(2) *International Journal for Court Administration* 30, 39–41.

68 Monika Zalnieriute, Lyria Bennett Moses and George Williams, 'The Rule of Law and Automation of Government Decision-Making' (2019) 82(3) *Modern Law Review* 425, 449.

69 Tania Sourdin, Bin Li and Tony Burke, 'Just Quick and Cheap? Civil Dispute Resolution and Technology' (2019) 19 *Macquarie Law Journal* 17, 23–4.

70 Richard M Re and Alicia Solow-Niederman, 'Developing Artificially Intelligent Justice' (2019) 22 *Stanford Technology Law Review* 242, 282–3.

71 Richard M Re and Alicia Solow-Niederman, 'Developing Artificially Intelligent Justice' (2019) 22 *Stanford Technology Law Review* 242, 282–3.

72 Richard M Re and Alicia Solow-Niederman, 'Developing Artificially Intelligent Justice' (2019) 22 *Stanford Technology Law Review* 242, 284.

73 Richard M Re and Alicia Solow-Niederman, 'Developing Artificially Intelligent Justice' (2019) 22 *Stanford Technology Law Review* 242, 285.

74 See Tania Sourdin, 'A Broader View of Justice?' in Michael Legg (ed), *Resolving Civil Disputes* (LexisNexis Australia, Chatswood, NSW, 2016), 19–36.

75 See for example, Adrian AS Zuckerman, 'Justice in Crisis: Comparative Dimensions of Civil Procedure' in Adrian AS Zuckerman (ed), *Civil Justice in Crisis* (Oxford University Press, 1999) 3; Adrian Zuckerman, *Zuckerman on Civil Procedure: Principles of Practice* (Sweet and Maxwell, 2nd ed, 2006).

76 See for example Jeremy Bentham, 'Principles of Judicial Procedure' in John Bowring (ed), *The Works of Jeremy Bentham Vol II* (William Tait, 1843) 5. For more modern views, see Justice Geoff Davies, 'The Reality of Civil Justice Reform: Why We Must Abandon the Essential Elements of Our System' (2003) 12 *Journal of Judicial Administration* 155; Nick Armstrong, 'Making Tracks' in Adrian Zuckerman and Ross Cranston (eds), *Reform of Civil Procedure: Essays on 'Access to Justice'* (Clarendon Press, 1995) 97; Murray Gleeson, 'The Judicial Method: Essentials and Inessentials' (2010) 9 *The Judicial Review* 377.

77 E Allan Lind and Tom R Tyler, *The Social Psychology of Procedural Justice* (Plenum Press, 1988).

78 See, for example, Kees Van den Bos, Lynn Van der Velden and Allan Lind, 'On the Role of Perceived Procedural Justice in Citizens' Reactions to Government Decisions and the Handling of Conflicts' (2014) 10 *Utrecht Law Review* 1; the base work of John Thibaut, 'Procedural Justice: A Psychological Analysis' (1978) 6 *Duke Law Journal* 1289; John Thibaut and Laurens Waler, *Procedural Justice: A Psychological Analysis* (Erlbaum, New Jersey, 1975).

79 See for example, Tania Sourdin, 'Dealing with Disputes about Tax in a "Fair" Way' (2015) 17(2) *Journal of Australian Taxation* 169, 222.

80 Tania Sourdin, 'The Role of the Courts in the New Justice System' (2015) 7 *Yearbook on Arbitration and Mediation* 98, 99.

81 Felicity Bell, 'Family Law, Access to Justice, and Automation' (2019) 19 *Macquarie Law Journal* 103, 128.

82 Allan E Barsky, 'The Ethics of App-Assisted Family Mediation' (2016) 34(1) *Conflict Resolution Quarterly* 31, 38.

83 Cited in Jeff Giddings and Michael Robertson, 'Informed Litigants with Nowhere to Go': Self-help Legal Aid Services in Australia' (2001) 26(4) *Alternative Law Journal* 184, 187.

84 Robert J Condlin, 'Online Dispute Resolution: Stinky, Repugnant, or Drab' (2017) 18(3) *Cardozo Journal of Conflict Resolution* 717, 721.

85 Robert J Condlin, 'Online Dispute Resolution: Stinky, Repugnant, or Drab' (2017) 18(3) *Cardozo Journal of Conflict Resolution* 717, 722.

86 Robert J Condlin, 'Online Dispute Resolution: Stinky, Repugnant, or Drab' (2017) 18(3) *Cardozo Journal of Conflict Resolution* 717, 721.

87 Richard M Re and Alicia Solow-Niederman, 'Developing Artificially Intelligent Justice' (2019) 22 *Stanford Technology Law Review* 242, 254.

88 Richard M Re and Alicia Solow-Niederman, 'Developing Artificially Intelligent Justice' (2019) 22 *Stanford Technology Law Review* 242, 255.

89 Richard M Re and Alicia Solow-Niederman, 'Developing Artificially Intelligent Justice' (2019) 22 *Stanford Technology Law Review* 242, 275–6.

90 Richard M Re and Alicia Solow-Niederman, 'Developing Artificially Intelligent Justice' (2019) 22 *Stanford Technology Law Review* 242, 276.

91 Richard M Re and Alicia Solow-Niederman, 'Developing Artificially Intelligent Justice' (2019) 22 *Stanford Technology Law Review* 242, 280.

92 Richard M Re and Alicia Solow-Niederman, 'Developing Artificially Intelligent Justice' (2019) 22 *Stanford Technology Law Review* 242, 280–1.

93 Richard M Re and Alicia Solow-Niederman, 'Developing Artificially Intelligent Justice' (2019) 22 *Stanford Technology Law Review* 242, 281.

94 See Tania Sourdin, Bin Li and Tony Burke, 'Just Quick and Cheap? Civil Dispute Resolution and Technology' (2019) 19 *Macquarie Law Journal* 17, 25.

95 Jena McGill, Suzanne Bouclin and Amy Salyzyn, 'Mobile and Web-based Legal Apps: Opportunities, Risks and Information Gaps' (2017) 15 *Canadian Journal of Law and Technology* 229, 244.

96 Jena McGill, Suzanne Bouclin and Amy Salyzyn, 'Mobile and Web-based Legal Apps: Opportunities, Risks and Information Gaps' (2017) 15 *Canadian Journal of Law and Technology* 229, 244.

97 Jena McGill, Suzanne Bouclin and Amy Salyzyn, 'Mobile and Web-based Legal Apps: Opportunities, Risks and Information Gaps' (2017) 15 *Canadian Journal of Law and Technology* 229, 244–5.

98 Teresa Scassa et al., 'Developing Privacy Best Practices for Direct-to-Public Legal Apps: Observations and Lessons Learned' (2020) 18(1) *Canadian Journal of Law and Technology* (forthcoming).

99 Jena McGill, Suzanne Bouclin and Amy Salyzyn, 'Mobile and Web-based Legal Apps: Opportunities, Risks and Information Gaps' (2017) 15 *Canadian Journal of Law and Technology* 229, 246.

100 Sherley Cruz, 'Coding for Cultural Competency: Expanding Access to Justice with Technology' (2019) 86 *Tennessee Law Review* 347, 367.

101 Sherley Cruz, 'Coding for Cultural Competency: Expanding Access to Justice with Technology' (2019) 86 *Tennessee Law Review* 347, 367.
102 Belinda Smyth and Bruce Fehlberg, 'Australian Post-Separation Parenting on the Smartphone: What's 'App-ening?' (2019) 41(1) *Journal of Social Welfare and Family Law* 53, 62.
103 Jane Bailey, Jacquelyn Burkell and Graham Reynolds, 'Access to Justice for All: Towards an 'Expansive Vision' of Justice and Technology' (2013) 31 *Windsor YB Access Just* 181, 197.
104 Teresa Scassa et al., 'Developing Privacy Best Practices for Direct-to-Public Legal Apps: Observations and Lessons Learned' (2020) 18(1) *Canadian Journal of Law and Technology* (forthcoming).
105 Noam Ebner and John Zeleznikow, 'Fairness, Trust and Security in Online Dispute Resolution' (2015) 36(2) *Hamline Journal of Public Law and Policy* 143, 154–6.
106 Noam Ebner and John Zeleznikow, 'Fairness, Trust and Security in Online Dispute Resolution' (2015) 36(2) *Hamline Journal of Public Law and Policy* 143, 158–9.
107 Noam Ebner and John Zeleznikow, 'Fairness, Trust and Security in Online Dispute Resolution' (2015) 36(2) *Hamline Journal of Public Law and Policy* 143, 155.
108 Teresa Scassa et al., 'Developing Privacy Best Practices for Direct-to-Public Legal Apps: Observations and Lessons Learned' (2020) 18(1) *Canadian Journal of Law and Technology* (forthcoming).

6 Future options

Introduction

As discussed in previous chapters, a wide range of justice apps that include mobile and web-based computer programmes, have been designed and developed to assist people in the justice sector. Considerable growth in connectivity, ODR processes, and the production and improvement of justice apps[1] have meant that justice apps have become an increasingly popular way of accessing information and connecting to services.[2] However, such apps have not been limited to the provision of information and legal advice and there are various notable examples of 'replacement' and 'disruptive' apps which are changing the way that dispute resolution and justice activities are being conducted around the world.

These developments are likely to expand in the near future as the justice system struggles to adjust to the COVID-19 pandemic which has not only led to severe disruptions to many courts but has also led to a greater need to remotely access justice services. In addition, the delays that were already present in respect of many court systems have been exacerbated by the pandemic which has led to the closure (or partial closure) of many courts and which may also lead to an increased demand for justice services.[3] The justice crisis caused by the pandemic may also mean that for some courts, future budgets will be cut, although others will receive some additional investment to ensure that technological supportive infrastructure can be constructed to support future activities. These changes are all likely to increase the need for, and adoption of, justice apps.

In some instances, the regulation and review of justice apps will be influenced by an overarching consideration of the app regulatory area. For example, some countries have begun to focus on creating more comprehensive responses towards the regulation of apps and the EU has, in particular, created frameworks that apply to apps more generally as part of establishing an ethical approach to the development of artificial intelligence (AI) and technology. This is doubtless an area of further future focus with the European Commission Justice Minister, Didier Reynders, having been specifically charged with responsibilities to not only maximise the potential of new digital technologies to improve the EU's justice systems, but to also contribute to a coordinated approach on the human and ethical implications of artificial intelligence.[4]

There are other factors that will influence the development of justice apps that are related to the differing cultural norms and approaches in different countries that can in turn be linked to how domestic justice systems are structured and used. For example, in China, as has been previously noted, justice apps may often be designed by a court, or by government agencies, or both. In addition, there may be a much higher level of acceptance of justice apps that use data from a range of sources, which would simply not be tolerated by citizens of other countries. For example, social surveillance data can be used in China to calculate a social credit score. The compilation may involve data from many agencies, and the score itself may be used to limit citizen movements and benefits, or in court proceedings, and could conceivably be used to assess credibility.[5]

In many instances, the applicability of regulatory arrangements and the review of justice apps will depend on the type of app, who created it, who uses it and other factors that can be linked to the type of data retained (and for how long). Essentially, in terms of regulatory and assessment approaches, a distinction must be drawn between apps that exist in the justice sector as they may vary according to whether:

(i) The app is not designed or intended to be used in the justice sector. Clearly many apps are not designed for justice sector use, however, such use may occur and in many circumstances can be beneficial. Examples of such apps could include *Zoom*, and *Microsoft Teams*.

(ii) The app is intended to be used in the justice sector and the users are identified as the public, court or other service users, a subset of such users (e.g., offenders), professional users (such as lawyers and mediators), or judges or justice sector staff (e.g., prosecution agencies, corrective services, parole officers, and others).

(iii) The app has a cost for users. For example, is the app intended to be 'free' and what might this mean? Is it assumed that data might be onsold or that advertising may take place or targeted referral? Is there an 'add on' upgrade that is sold commercially? Is the app available on a subscription plan or through some other payment service (e.g., as an 'add on' through another subscription)?

(iv) The app is created or managed by courts, judges, a government agency, law firm, legal centre, not for profit (e.g., a University), private or commercial interest, or some combination. Additional ethical protections may, for example, be required for some apps depending on the nature and identity of the creator or manager.

(v) The clarity of purpose of the app. For example, whether the app is intended to be supportive, replacement or disruptive, or what combination of these three types of technologies is envisaged or developed. The authors note that the *Adieu* app, for example, many have initially been intended to focus on the first two types of technologies, however, the creation of an AI powered financially focussed app might mean that all three types of technology are present.

These five points are relevant as they enable evaluation of justice apps to take place and because they also raise issues in terms of the capacity of apps to improve the justice sector. The four stage framework developed in Chapter 3 and explored in Chapter 5 also remains relevant, although, the weight or emphasis given to each sub-objective within each factor may vary. Essentially, that framework requires consideration of four factors: (i) ease of use; (ii) effectiveness; (iii) privacy and security; and (iv) interoperability. By way of example, effectiveness requires consideration of justice objectives, which may require a greater focus depending on the characteristics of the app.

Other factors that will continue to influence app development are linked to broader regulatory arrangements and the way that users might work with apps. Some factors are linked to cultural impediments, such as a reluctance to innovate (see discussion in Chapter 2). As apps become more focussed on complex advice arrangements, some particular issues have emerged relating to how and to what extent legal services can be linked to and related to justice apps. An example of these issues is explored below and relates to the way in which lawyers may provide unbundled legal services or exercise an oversight function in terms of the operation of an app.

Jurisdictional issues with apps – the unbundling of legal services

As noted above, some apps raise particular issues because of the jurisdictions within which they operate. For example, in a number of jurisdictions, there are issues about the extent to which an app can support or even replace the giving of legal advice by a human. Whilst many justice apps may not be directed at legal advice, this remains an increasingly important area of justice app development.

In a number of jurisdictions, there can be legal and ethical issues associated with the delivery of legal services by non-lawyers. This is known as the 'unauthorised practice of law' and is illegal.[6] For example, in Australia, s 10 of the *Legal Profession Uniform Law* (NSW) prohibits unqualified entities from engaging in legal practice. McGill, Bouclin, and Salyzyn explain that whilst non-lawyers cannot deliver legal services, they can provide legal information. However, the authors note that 'the line between legal information and legal services is notoriously murky'.[7] Also in Australia, Waye, Verreynne, and Knowler have noted that 'as technological advances permit customisable legal information and documentation to be delivered at scale, like the line between legal and managerial matters ... the line between the provision of legal information and the provision of legal advice is now much harder to draw'.[8]

According to the Productivity Commission of Australia, 'non-legal professionals have, for some time, been providing advice (and in some cases advocacy) in a range of areas including conveyancing, intellectual property, workplace relations, taxation and migration'.[9]

In attempting to draw a line between legal information and legal advice, Giddings and Robertson have noted that whereas legal information relates to 'generic information which does not address individual circumstances', legal advice 'is taken as being tailored to the individual circumstances of the client'.[10] Cognisant of this distinction, self-help kits – such as those published by Legal Aid Queensland which cover domestic violence and responsibility for children post-separation – contain a disclaimer that the material provided is intended as 'information' only, with users who have a 'legal problem' directed to seek advice from a lawyer.[11]

Lupica, Franklin, and Friedman have also considered whether technology-based self-help tools implicate (or should implicate) rules prohibiting the unauthorised practice of law.[12] The authors refer to a ruling of the New York Court of Appeal that where there is no 'personal contact or relationship with a particular individual', nor the 'relation of confidence of trust so necessary to the status of attorney and client', there is no 'practice of law'.[13] However, it is also noted that justice apps are considerably more interactive than other types of self-help tools such as books.[14] Significantly, it is argued that 'as technology becomes more advanced, computer programmes will make greater use of artificial intelligence and become more "human-like" in their problem-solving ability'.[15] Lupica et al. conclude by noting that if justice apps raise issues related to the unauthorised practice of law, the legal profession 'must reconsider the definition of the unauthorised practice of law to account for creative solutions being developed to expand access to civil justice'.[16]

The issue of unauthorised practice of law also exists in the broader ODR context and particularly in the e-commerce setting where some more innovative approaches have been adopted that exclude lawyers altogether. *Taobao*, for example – a Chinese online shopping website owned by Alibaba and the world's biggest e-commerce website – has developed its own ODR system by which buyers and sellers can settle their commercial disputes. *Taobao*'s ODR system consists of three stages, being online negotiation, public assessment, and *Taobao*'s decision. When a dispute arises between a buyer and seller, *Taobao* provides an online platform for the two parties to negotiate, with the aim of settling the dispute. If the negotiation is unsuccessful, the buyer then has an option to proceed to the public assessment stage, or to ask *Taobao* staff to deliver a decision straight away. Public assessment is a process where 31 volunteers are randomly selected from a list kept by *Taobao* to form an online 'jury' to hear the 'case' and assess the relevant 'evidence'. The party gaining at least 16 votes out of 31 volunteers wins the case. Any persons, subject to their social credit records, can apply to become volunteers for this public assessment system. If neither party gains at least 16 votes, then *Taobao* staff will step in to make a decision about the dispute as the last step of this ODR system.[17]

According to a report released by Alibaba in January 2019, as of December 2018, over 15 million disputes had been successfully addressed through the participation of around 3 million volunteers in public assessment.[18] Despite this significant achievement, Qi and Huang have identified several challenges facing this ODR system that may adversely impact on a fair outcome, including the possibility of increased delay if parties do not accept the decision of the public assessors or *Taobao* staff, and the fact that assessors are not legal professionals. Qi and Huang further suggest that there is a need to link *Taobao*'s ODR system to an e-commerce tribunal.[19]

Issues surrounding the extent and nature of any lawyer involvement in terms of overview functions have also arisen in relation to smart contracts. The legality of such contracts is a live question in the literature, with Arachchi noting that jurisprudential questions remain as to how smart contracts are to be characterised.[20] A 2016 report by R3 and Norton Rose Fulbright concluded that using the term 'contract' does not automatically render its contents binding, noting that the electronic nature of the document causes issues of validity.[21] Williams also observes that as payments, asset transfers or other contractual obligations may be automatically triggered in smart contracts, it remains unclear how traditional contractual principles and enforcement mechanisms would apply.[22]

In a number of countries such as Australia, there has also been some consideration of these and related issues in the context of unbundled legal services. Castles explains that the unbundling of legal services, also known as 'discrete task assistance' or 'step-in-step-out' legal representation, works on the basis that 'lawyers are retained to do some, but not all, of the transactional or litigious legal work for their client, with the client shouldering responsibility for the remainder'.[23] The Productivity Commission has described unbundling as a 'half-way house' between full representation and no representation.[24] Variables determining the type and depth of services that clients may seek from a solicitor providing unbundled services include: the extent and accuracy of information given to the client, the client's personality, the complexity of the task, and the costs and resources available to do the job.[25]

The key benefit of unbundling is that it can help facilitate access to justice by bridging the gap between the high cost of legal representation and diminishing legal aid resources.[26] As noted by Castles, this can be of particular assistance to the 'missing middle' – 'the demographic with steady income, who nonetheless cannot hope to afford the full cost of defending or prosecuting a litigious matter in court'.[27] In its 2014 *Access to Justice Arrangements* report, the Productivity Commission, in summarising the key policy reasons for unbundling, referred to the market benefit of empowering consumers by enabling them to select services to best meet their needs.[28]

It has also been noted that unbundling can assist the growing number of self-represented litigants.[29] Further, a number of commentators have specifically highlighted the role that justice apps can play in closing the justice gap,[30] with Brown highlighting the importance of striking a balance between regulation and accessibility.[31] In addition to the high costs of legal representation, Mosten has identified three other reasons for the unbundling of legal services: excessive delays from congested courts, defensive lawyering, and 'the acrimony caused by the jousting of the adversarial system'.[32]

Whilst unbundled legal service arrangements are more common in a number of jurisdictions – such as the U.S., Canada, and the UK – where there exist detailed rules, and ethical and practice guidelines relating to the provision of such services[33] (see however the discussion below in relation to the creation of new entities to remove some regulatory hurdles), in Australia, the approach has been much more cautious. There are no formal rules for the provision of unbundled legal services,[34] and although existing rules do not explicitly preclude the delivery of such services, they may provide a disincentive to the private practice providing unbundled legal assistance.[35] As outlined by Castles, court rules, ethical guidelines and legal practice models are all based on the traditional assumption that litigation will be conducted, from start to finish, through a lawyer.[36]

The Law Society of England and Wales has also referred to the risk of a poorer service being offered by solicitors in a less vigorous way.[37] A further risk of unbundling, noted by Castles, relates to the 'grey areas' outside of the written terms of a solicitor's retainer that may be imputed to the contractual terms or superimposed as a tortious duty. If a lawyer is working in a 'step-in-step-out' capacity and is not familiar with the whole case history and/or background, it will be much more difficult to foresee these hidden duties.[38] Giddings and Robertson have similarly noted that legal service providers engaged in unbundling can face issues related to the provision of incorrect advice where their limited retainer prevents them from researching, analysing, and reflecting on the complexity of the legal problem.[39]

Issues of this nature were considered in the UK by the Court of Appeal in 2015, in a case of alleged professional negligence in the context of unbundled legal services.[40] The Court held that a solicitor on a retainer to redraft financial orders in a family law matter had no duty to advise on the underlying financial agreement, noting that good practice requires limiting the scope of the retainer in writing. Finally, when it comes to the practical challenges associated with unbundling, there may be certain tasks which are unable to be divided into separate components capable of being unbundled.[41]

In light of such risks, unbundling guidelines in some jurisdictions explicitly note that some cases are not suitable for unbundled services, meaning some representation is not always better than no representation.[42] Nevertheless, the Productivity Commission of Australia has concluded that barriers to unbundling can be overcome, recommending changes to the Australian Solicitors Conduct Rules to provide greater clarity and certainty around this practice.[43] Further, the Commission has noted the need for professional associations to improve awareness of unbundling among their members through guidance documents.[44] Legg has similarly argued that legal professional legislation and/or ethics or court rules must be revisited so as to support the provision of unbundled legal services and provide guidance on how and when such services should operate.[45] It is noted that 'the confluence of concerns about the affordability of legal services and the greater use of technology to provide legal information and related services' means that the number of clients seeking limited scope services will only continue to grow.[46]

Whilst in the past, the use of technology to deliver unbundled services of this nature was opposed on the basis of the 'digital divide',[47] as noted in Chapter 5, the 'digital divide' has decreased as simpler technologies have evolved, internet access has increased across communities, and technological competencies and preferences have grown.[48] Kimbro has summarised some of the ways in which technology has been used to facilitate the unbundling of legal services.[49] These include:

(i) The use of document-assembly and automation tools that enable information to be collected from clients online;

(ii) Decision-making and AI tools that assist lawyers to generate reports to guide an unbundled client, or which the client themselves may access to receive instruction for self-help representation;

(iii) Online case and client management to streamline the process of working with unbundled clients by keeping digital records of the legal documents and/or guidance and instruction that has been provided to the client;

(iv) Online Branded Networks that allow lawyers to deliver unbundled legal services online directly to consumers (e.g., *Rocket Lawyer*);

(v) Web calculators and advisors which directly appeal to prospective self-help clients and which may then guide the client to turn to the organisation for unbundled or full services;

(vi) Guided walkthrough tools which allow self-help clients to navigate through a series of questions to determine whether or not they have a legal need, and provide information on how to address it;

(vii) Real-time or 'live help' chat lawyers which can engage in real-time chat using free or subscription-based services; and

(viii) Short video tutorials or podcasts posted online.[50]

There is some uncertainty surrounding the difficult issue of whether, and in what circumstances, online platforms which provide unbundled legal services breach rules against the unauthorised practice of law which are summarised above. Waye, Verreynne, and Knowler identify *LegalZoom* – an online American company that operates in Australia – as a case in point.[51] *LegalZoom* offers personalised downloadable legal documentation such as wills and trademark applications.[52] As outlined by Figueras, *LegalZoom* is threatened by the unauthorised practice of law doctrine because part of its service involves preparing legal documents for clients based on their responses to questionnaires, as opposed to simply providing blank forms.[53] Put simply, *LegalZoom* crosses the line from form provider to form preparer.[54] This has seen the company's operations challenged through a number of investigations by State Bar Associations, lawsuits, and cease-and-desist letters,[55] with violations of unauthorised practice of law rules established in at least some jurisdictions.[56] Waye, Verreynne and Knowler have noted that the Australian position in respect of platforms such as *LegalZoom* is unclear.[57] On the one hand, the provision of such legal documentation appears to be caught by the prohibition on unqualified entities engaging in legal practice. On the other hand, however:

> Australian courts draw a distinction between the provision of generic legal documentation for completion by clients, and the provision of customised documentation that requires careful drafting by a legally qualified expert or which is accompanied by the provision of personalised legal advice. The former is not categorised as engaging in legal practice, whereas the latter is prohibited without legal credentialing. The distinction is regarded as a matter of fact and degree rather than a bright line rule.[58]

Simshaw has noted that these same issues around the unauthorised practice of law arise in relation to *DoNotPay* – an AI chatbot app which is available in the U.S. and the UK.[59]

The uncertainty surrounding the interaction between the provision of unbundled legal services and unauthorised practice of law rules has been identified as problematic. Beyond the access to justice considerations canvassed above, Scassa et al. have noted that the possibility of breaching unauthorised practice of law rules creates a 'chilling effect on innovation'.[60] Specifically, it is noted that while 'large actors like *LegalZoom* may be willing to take on potential regulatory investigations and legal proceedings as acceptable risks of doing business', many smaller direct-to-public justice app developers have 'expressed significant caution and concern' and may be unlikely to take on such risks.[61]

In the USA there is some experimentation currently underway to enable various regulatory hurdles to be addressed that might prevent the unbundling of legal services or the creation of new entities that may support justice app development. For example, the California Task Force on Access through Innovation of Legal Services[62] recommended in early 2020 that a regulatory sandbox be set up that would permit entities to be created to provide legal services that would previously have been prohibited under California Rules of Professional Responsibility. Similarly, the Institute for the Advancement of the American Legal System (IAALS) has a sandbox project that is focussed on unlocking lawyer's regulatory arrangements.[63]

Finally, the family law context also raises unique issues when it comes to the provision of unbundled legal services. In its 2017 Guidance Statement on limited scope representation, the Queensland Law Society in Australia noted that when determining if a matter is suitable for limited scope representation, factors such as the level of conflict in the dispute must be considered. It is noted that 'cases where there is a history of entrenched conflict between the parties pose obvious challenges for limited scope assistance'.[64]

Conclusions

Despite the growth in ODR and justice apps, issues remain that are associated with the use of technology in the justice sector, including a general reluctance to innovate, particularly in the private legal practice area.[65] This is unfortunate because justice apps show considerable promise in terms of improving the justice sector. First and foremost, new technologies of this nature can create new pathways to justice.[66] For instance, justice apps can assist with triage and referral functions, essentially enabling people to be referred to a human expert or in some cases a well-developed chatbot so that a justice issue is attended to as soon as possible. The creation of new pathways to justice is particularly important in the family law sector where access to justice has been identified as a serious problem.[67] This is also the case in the criminal law sector where apps could help defendants navigate the complex justice system and even contribute to justice reform.

In addition to technology reducing cost and delay, it is also capable of mitigating psychological, informational, and physical barriers.[68] At the same time, however, there are issues with embracing all justice apps where they may result in the dehumanisation of justice processes, that in turn results in the system becoming less 'just'.[69] It is also suggested that levels of reluctance vary across jurisdictions. In China, for example, technology seems to be more widely employed in the court system and used to assist with judicial work than in courts in other countries. This is probably driven by the huge case volumes presented by China's large population, as well as very different understandings about the role of judges and the executive in the administration of the justice system.[70]

In the family law context, justice apps can also enable parties to avoid an 'in room' face-to-face interaction and reduce costs,[71] although research suggests that apps may never be able to completely replace interaction with a lawyer.[72] In particular, automated options should not be viewed as appropriate substitutes for professional family lawyers where there are vulnerable clients and children involved.[73]

Justice app use must be considered in light of the various issues and limitations associated with app usage in the justice sector. These challenges can be attended to be considering the evaluation framework noted in Chapter 3 of this book and earlier in this chapter. Specifically, challenges vary according to the nature of the app being considered and for newer apps that may incorporate advisory and determinative approaches, there are additional challenges posed by the digital divide, accessibility issues, issues associated with the translation of law into code, the challenging question of whether automated systems are appropriate in the context of discretionary decision-making and value judgments, various justice, privacy, security, and confidentially considerations, and issues associated with the unauthorised practice of law and the unbundling of legal services. To ensure that justice apps support justice, they may require the creation of new regulatory frameworks that apply to lawyers and which encourage innovation. In addition, justice apps require human expert planning input, trials, and evaluation using an appropriate framework, with emphasis on the various components discussed in this book, and some consideration of the cultural factors that may lead to a reluctance to innovate in parts of the justice system.

Notes

1 See, e.g., Australian Bureau of Statistics, *Household Use of Information Technology, Australia, 2016–17* (Catalogue No 8146.0, 28 March 2018); Tania Sourdin and Chinthaka Liyanage, 'The Promise and Reality of Online Dispute Resolution in Australia' in Mohamed S Abdel Wahab, Ethan Katsh and Daniel Rainey (eds), *Online Dispute Resolution: Theory and Practice a Treatise on Technology and Dispute Resolution* (Eleven International Publishing, 2012) 483, 483.
2 See generally Tania Sourdin and Chinthaka Liyanage, 'The Promise and Reality of Online Dispute Resolution in Australia' in Mohamed S Abdel Wahab, Ethan Katsh and Daniel Rainey (eds), *Online Dispute Resolution: Theory and Practice a Treatise on Technology and Dispute Resolution* (Eleven International Publishing, 2012) 483, 484–5.
3 Tania Sourdin and John Zeleznikow, 'Courts, Mediation and COVID-19', *Australian Business Law Review* 48 (2020) 138.

4 See <https://ec.europa.eu/commission/commissioners/2019-2024/reynders_en>.

5 See Tania Sourdin, *Judges, Technology and AI*, forthcoming (2021) (Edward Elgar); see also Rogier Creemers, 'China's Social Credit System: An Evolving Practice of Control' (May 9, 2018). Available at SSRN: <https://ssrn.com/abstract=3175792> and for an example of how this is perceived in China see Dev Lewis, 'Separating Myth from Reality, How China's Social Credit System Uses Public Data for Social Governance' Monday, 18 May 2020 available at <https://www.nesta.org.uk/report/separating-myth-reality/citizen-scoring/>.

6 See generally Jena McGill, Suzanne Bouclin and Amy Salyzyn, 'Mobile and Web-based Legal Apps: Opportunities, Risks and Information Gaps' (2017) 15 *Canadian Journal of Law and Technology* 229, 248.

7 Jena McGill, Suzanne Bouclin and Amy Salyzyn, 'Mobile and Web-based Legal Apps: Opportunities, Risks and Information Gaps' (2017) 15 *Canadian Journal of Law and Technology* 229, 249.

8 Vicki Waye, Martie-Louise Verreynne and Jane Knowler, 'Innovation in the Australian Legal Profession' (2018) 25(2) *International Journal of the Legal Profession* 213, 219–20.

9 Productivity Commission of Australia, *Access to Justice Arrangements* (Report No. 72, 5 September 2014) 21.

10 Jeff Giddings and Michael Robertson, 'Informed Litigants with Nowhere to Go': Self-help Legal Aid Services in Australia' (2001) 26(4) *Alternative Law Journal* 184, 187.

11 Jeff Giddings and Michael Robertson, 'Informed Litigants with Nowhere to Go': Self-help Legal Aid Services in Australia' (2001) 26(4) *Alternative Law Journal* 184, 187.

12 Lois R Lupica, Tobias A Franklin and Sage M Friedman, 'The Apps for Justice Project: Employing Design Thinking to Narrow the Access to Justice Gap' (2017) 44(5) *Fordham Urban Law Journal* 1363.

13 Lois R Lupica, Tobias A Franklin and Sage M Friedman, 'The Apps for Justice Project: Employing Design Thinking to Narrow the Access to Justice Gap' (2017) 44(5) *Fordham Urban Law Journal* 1363, 1401.

14 Lois R Lupica, Tobias A Franklin and Sage M Friedman, 'The Apps for Justice Project: Employing Design Thinking to Narrow the Access to Justice Gap' (2017) 44(5) *Fordham Urban Law Journal* 1363, 1401.

15 Lois R Lupica, Tobias A Franklin and Sage M Friedman, 'The Apps for Justice Project: Employing Design Thinking to Narrow the Access to Justice Gap' (2017) 44(5) *Fordham Urban Law Journal* 1363, 1402.

16 Lois R Lupica, Tobias A Franklin and Sage M Friedman, 'The Apps for Justice Project: Employing Design Thinking to Narrow the Access to Justice Gap' (2017) 44(5) *Fordham Urban Law Journal* 1363, 1402.

17 See Alibaba, *The Inheritance and Development of Fengqiao Experience in Alibaba* (2019), 2 <http://www.xinhuanet.com/tech/2019-01/08/c_1123959245.htm>.

18 See Alibaba, *The Inheritance and Development of Fengqiao Experience in Alibaba* (2019), 2 <http://www.xinhuanet.com/tech/2019-01/08/c_1123959245.htm>.

19 Zhilong Qi, Muchen Huang, 'The Docking between Taobao ODR System and An E-commerce Tribunal', *Science and Technology Information* (2019), <http://m.fx361.com/news/2019/1210/6163096.html>.

20　Buwaneka Arachchi, 'Chains, Coins and Contract Law: The Validity and Enforceability of Smart Contracts' (2019) 47 *Australian Business Law Review* 40, 41.

21　R3 and Norton Rose Fulbright, 'Can Smart Contracts Be Legally Binding Contracts?' (White Paper, Norton Rose Fulbright, November 2016) <https://www.nortonrosefulbright.com/-/media/files/nrf/nrfweb/imported/norton-rose-fulbright--r3-smart-contracts-white-paper-key-findings-nov-2016.pdf>.

22　Michael Williams, "Lawyers, Technology and Dispute Resolution" in Michael Legg (ed), *Resolving Civil Disputes*, Lexisnexis Butterworths (2016), 350–1.

23　Margaret Castles, 'Barriers to Unbundled Legal Services in Australia: Canvassing Reforms to Better Manage Self-Represented Litigants in Courts and in Practice' (2016) 25(4) *Journal of Judicial Administration* 237, 237.

24　Productivity Commission of Australia, *Access to Justice Arrangements* (Report No. 72, 5 September 2014) 20. See also Stephanie Kimbro, 'Using Technology to Unbundle in the Legal Services Community', *Harvard Journal of Law & Technology Occasional Paper Series* (online, February 2013) 1, 1.

25　Forrest S Mosten, 'Unbundling of Legal Services and the Family Lawyer' (1994) 28(3) *Family Law Quarterly* 421, 423.

26　Margaret Castles, 'Barriers to Unbundled Legal Services in Australia: Canvassing Reforms to Better Manage Self-Represented Litigants in Courts and in Practice' (2016) 25(4) *Journal of Judicial Administration* 237, 237; Stephanie Kimbro, 'Using Technology to Unbundle in the Legal Services Community', *Harvard Journal of Law & Technology Occasional Paper Series* (online, February 2013) 1, 1; Michael Legg, 'Recognising a New Form of Legal Practice: Limited Scope Services' (2018) 50 *Law Society of NSW Journal* 74; Queensland Law Society, *Guidance Statement No. 7 – Limited Scope Representation in Dispute Resolution* (8 June 2017) <https://www.qls.com.au/Knowledge_centre/Ethics/Guidance_Statements/Guidance_Statement_No_7_-_Limited_scope_representation_in_dispute_resolution> 1.

27　Margaret Castles, 'Barriers to Unbundled Legal Services in Australia: Canvassing Reforms to Better Manage Self-Represented Litigants in Courts and in Practice' (2016) 25(4) *Journal of Judicial Administration* 237, 238.

28　Productivity Commission of Australia, *Access to Justice Arrangements* (Report No. 72, 5 September 2014) 2.

29　Margaret Castles, 'Barriers to Unbundled Legal Services in Australia: Canvassing Reforms to Better Manage Self-Represented Litigants in Courts and in Practice' (2016) 25(4) *Journal of Judicial Administration* 237, 240.

30　See, eg, Isaac Figueras, 'The LegalZoom Identity Crisis: Legal Form Provider or Lawyer in Sheep's Clothing?' (2013) 63(4) *Case Western Reserve Law Review* 1419, 1421; Caroline E Brown, 'LegalZoom: Closing the Justice Gap or Unauthorized Practice of Law?' (2016) 17(5) *North Carolina Journal of Law & Technology* 219, 248.

31　Caroline E Brown, 'LegalZoom: Closing the Justice Gap or Unauthorized Practice of Law?' (2016) 17(5) *North Carolina Journal of Law & Technology* 219, 247.

32 Forrest S Mosten, 'Unbundling of Legal Services and the Family Lawyer' (1994) 28(3) *Family Law Quarterly* 421, 424–5.

33 See generally Margaret Castles, 'Barriers to Unbundled Legal Services in Australia: Canvassing Reforms to Better Manage Self-Represented Litigants in Courts and in Practice' (2016) 25(4) *Journal of Judicial Administration* 237, 238.

34 Michael Legg, 'Recognising a New Form of Legal Practice: Limited Scope Services' (2018) 50 *Law Society of NSW Journal* 74; Margaret Castles, 'Barriers to Unbundled Legal Services in Australia: Canvassing Reforms to Better Manage Self-Represented Litigants in Courts and in Practice' (2016) 25(4) *Journal of Judicial Administration* 237, 239.

35 Justice Connect, 'Unbundling and the "Missing Middle": Submission to the Law Council of Australia's Review of the Australian Solicitor's Conduct Rules' (July 2018) 7.

36 Margaret Castles, 'Barriers to Unbundled Legal Services in Australia: Canvassing Reforms to Better Manage Self-Represented Litigants in Courts and in Practice' (2016) 25(4) *Journal of Judicial Administration* 237, 240.

37 Law Society of England and Wales, *Lawyers Respond to Legal Aid Crisis by Unbundling Legal Services for Clients* (Press Release, 1 May 2013) [3.1], cited in Margaret Castles, 'Barriers to Unbundled Legal Services in Australia: Canvassing Reforms to Better Manage Self-Represented Litigants in Courts and in Practice' (2016) 25(4) *Journal of Judicial Administration* 237, 240.

38 Margaret Castles, 'Barriers to Unbundled Legal Services in Australia: Canvassing Reforms to Better Manage Self-Represented Litigants in Courts and in Practice' (2016) 25(4) *Journal of Judicial Administration* 237, 253.

39 Jeff Giddings and Michael Robertson, 'Informed Litigants with Nowhere to Go': Self-help Legal Aid Services in Australia' (2001) 26(4) *Alternative Law Journal* 184, 189.

40 *Minkin v Landsberg* [2015] EWCA Civ 1152, [43].

41 Jeff Giddings and Michael Robertson, 'Informed Litigants with Nowhere to Go': Self-help Legal Aid Services in Australia' (2001) 26(4) *Alternative Law Journal* 184, 189.

42 See generally Margaret Castles, 'Barriers to Unbundled Legal Services in Australia: Canvassing Reforms to Better Manage Self-Represented Litigants in Courts and in Practice' (2016) 25(4) *Journal of Judicial Administration* 237, 253.

43 Productivity Commission of Australia, *Access to Justice Arrangements* (Report No. 72, 5 September 2014) 650.

44 Productivity Commission of Australia, *Access to Justice Arrangements* (Report No. 72, 5 September 2014) 650.

45 Michael Legg, 'Recognising a New Form of Legal Practice: Limited Scope Services' (2018) 50 *Law Society of NSW Journal* 74.

46 Michael Legg, 'Recognising a New Form of Legal Practice: Limited Scope Services' (2018) 50 *Law Society of NSW Journal* 74, 97.

47 See generally Stephanie Kimbro, 'Using Technology to Unbundle in the Legal Services Community', *Harvard Journal of Law & Technology Occasional Paper Series* (online, February 2013) 1, 18.

48 Tania Sourdin, Bin Li and Tony Burke, 'Just Quick and Cheap? Civil Dispute Resolution and Technology' (2019) 19 *Macquarie Law Journal* 17, 36.

49 Stephanie Kimbro, 'Using Technology to Unbundle in the Legal Services Community', *Harvard Journal of Law & Technology Occasional Paper Series* (online, February 2013) 1, 18–23.

50 This material is drawn from and discussed in more detail in Stephanie Kimbro, 'Using Technology to Unbundle in the Legal Services Community', *Harvard Journal of Law & Technology Occasional Paper Series* (online, February 2013) 1, 18–23.

51 Vicki Waye, Martie-Louise Verreynne and Jane Knowler, 'Innovation in the Australian Legal Profession' (2018) 25(2) *International Journal of the Legal Profession* 213, 220.

52 LegalZoom, *LegalZoom* <https://www.legalzoom.com/country/au>.

53 Isaac Figueras, 'The LegalZoom Identity Crisis: Legal Form Provider or Lawyer in Sheep's Clothing?' (2013) 63(4) *Case Western Reserve Law Review* 1419, 1429.

54 Isaac Figueras, 'The LegalZoom Identity Crisis: Legal Form Provider or Lawyer in Sheep's Clothing?' (2013) 63(4) *Case Western Reserve Law Review* 1419, 1429.

55 Teresa Scassa et al., 'Developing Privacy Best Practices for Direct-to-Public Legal Apps: Observations and Lessons Learned' (2020) 18(1) *Canadian Journal of Law and Technology* (forthcoming).

56 Isaac Figueras, 'The LegalZoom Identity Crisis: Legal Form Provider or Lawyer in Sheep's Clothing?' (2013) 63(4) *Case Western Reserve Law Review* 1419, 1430.

57 Vicki Waye, Martie-Louise Verreynne and Jane Knowler, 'Innovation in the Australian Legal Profession' (2018) 25(2) *International Journal of the Legal Profession* 213, 220.

58 Vicki Waye, Martie-Louise Verreynne and Jane Knowler, 'Innovation in the Australian Legal Profession' (2018) 25(2) *International Journal of the Legal Profession* 213, 220, citing *Legal Practice Board v Giraudo* [2010] WASC 4 [12]–[13].

59 Drew Simshaw, 'Ethical Issues in Robo-Lawyering: The Need for Guidance on Developing and Using Artificial Intelligence in the Practice of Law' (2018) 70 *Hastings Law Journal* 173, 178.

60 Teresa Scassa et al., 'Developing Privacy Best Practices for Direct-to-Public Legal Apps: Observations and Lessons Learned' (2020) 18(1) *Canadian Journal of Law and Technology* (forthcoming).

61 Teresa Scassa et al., 'Developing Privacy Best Practices for Direct-to-Public Legal Apps: Observations and Lessons Learned' (2020) 18(1) *Canadian Journal of Law and Technology* (forthcoming).

62 See https://www.calbar.ca.gov/About-Us/Who-We-Are/Committees/Task-Force-on-Access-Through-Innovation-of-Legal-Services.

63 See material on this project at <https://iaals.du.edu/projects/unlocking-legal-regulation>.

64 Queensland Law Society, *Guidance Statement No. 7 – Limited Scope Representation in Dispute Resolution* (8 June 2017) <https://www.qls.com.au/Knowledge_centre/Ethics/Guidance_Statements/Guidance_Statement_No_7_-_Limited_scope_representation_in_dispute_resolution> 4.

65 Tania Sourdin, Bin Li and Tony Burke, 'Just Quick and Cheap? Civil Dispute Resolution and Technology' (2019) 19 *Macquarie Law Journal* 17, 32.

66 See, eg, Lisa Toohey et al., 'Meeting the Access to Civil Justice Challenge: Digital Inclusion, Algorithmic Justice, and Human-Centred Design' (2019) 19 *Macquarie Law Journal* 133, 143; Sourdin, Tania, Bin Li and Tony Burke, 'Just Quick and Cheap? Civil Dispute Resolution and Technology' (2019) 19 *Macquarie Law Journal* 17, 18; Melissa Conley Tyler and Mark McPherson, 'Online Dispute Resolution and Family Disputes' (2006) 12(2) *Journal of Family Studies* 165; David Luban, 'Optimism, Skepticism and Access to Justice' (2016) 3(3) *Texas A&M Law Review* 495, 502; Jessica Frank, 'A2J Author, Legal Aid Organizations, and Courts: Bridging the Civil Justice Gap Using Document Assembly' (2017) 39(2) *Western New England Law Review* 251.

67 Felicity Bell, 'Family Law, Access to Justice, and Automation' (2019) 19 *Macquarie Law Journal* 103, 113; Chief Justice Alastair Nicholson, 'Legal Aid and a Fair Family Law System' (Legal Aid Forum Towards 2010, Australian Capital Territory, 21 April 1999).

68 Jena McGill, Suzanne Bouclin and Amy Salyzyn, 'Mobile and Web-based Legal Apps: Opportunities, Risks and Information Gaps' (2017) 15 *Canadian Journal of Law and Technology* 229, 241–3.

69 Tania Sourdin, Bin Li and Tony Burke, 'Just Quick and Cheap? Civil Dispute Resolution and Technology' (2019) 19 *Macquarie Law Journal* 17, 18.

70 According to *the Work Report of the Supreme People's Court in 2018*, Chief Justice Qiang Zhou, President of SPC, noted that there were more than 28 million cases filed across the country in the year of 2018 alone. This has led to huge workload for Chinese judges. SPC noted that in the year of 2017, the average closed case numbers per judge per year in Zhejiang Province was 315, ranked No.1 in the country.

71 Melissa Conley Tyler and Mark McPherson, 'Online Dispute Resolution and Family Disputes' (2006) 12(2) *Journal of Family Studies* 165, 170; Tania Sourdin and Chinthaka Liyanage, 'The Promise and Reality of Online Dispute Resolution in Australia' in Mohamed S Abdel Wahab, Ethan Katsh and Daniel Rainey (eds), *Online Dispute Resolution: Theory and Practice a Treatise on Technology and Dispute Resolution* (Eleven International Publishing, 2012) 483, 499; Felicity Bell, 'Family Law, Access to Justice, and Automation' (2019) 19 *Macquarie Law Journal* 103, 119; Emilia Bellucci, Sitalakshmi Venkatraman and Andrew Stranieri, 'Online Dispute Resolution in Mediating EHR Disputes: A Case Study on the Impact of Emotional Intelligence' (2019) *Behaviour & Information Technology* 3, 6.

72 Lyria Bennett Moses, 'Artificial Intelligence in the Courts, Legal Academia and Legal Practice' (2017) 91(7) *Australian Law Journal* 561, 567–8; Felicity Bell, 'Family Law, Access to Justice, and Automation' (2019) 19 *Macquarie Law Journal* 103, 131–2.

73 Dana Remus and Frank Levy, 'Can Robots Be Lawyers: Computers, Lawyers, and the Practice of Law' (2017) 30(3) *Georgetown Journal of Legal Ethics* 501.

Index

Note: Page numbers followed by "n" denote endnotes.

Printed in the United States
by Baker & Taylor Publisher Services

Printed in the United States
by Baker & Taylor Publisher Services